WORKSHEETS
FOR CLASSROOM OR LAB PRACTICE

CARRIE GREEN

BEGINNING & INTERMEDIATE ALGEBRA: BUILDING A FOUNDATION

Paul McKenna
San Antonio College

Helen Kirk
Palo Alto College

Addison-Wesley
is an imprint of

PEARSON

The author and publisher of this book have used their best efforts in preparing this book. These efforts include the development, research, and testing of the theories and programs to determine their effectiveness. The author and publisher make no warranty of any kind, expressed or implied, with regard to these programs or the documentation contained in this book. The author and publisher shall not be liable in any event for incidental or consequential damages in connection with, or arising out of, the furnishing, performance, or use of these programs.

Reproduced by Pearson Addison-Wesley from electronic files supplied by the author.

Publishing as Pearson Addison-Wesley, 75 Arlington Street, Boston, MA 02116.

ISBN-13: 978-0-321-59397-9
ISBN-10: 0-321-59397-9

1 2 3 4 5 6 OPM 12 11 10 09

Addison-Wesley
is an imprint of

PEARSON

www.pearsonhighered.com

Table of Contents

Dear Instructors and Students,

The **Worksheets for Classroom or Lab Practice** is a workbook that was designed to be used in a variety of ways. We see it as an organizational tool; a self- and teacher-assessment tool; a place to get more practice, to show work, and to keep notes. Every section of *Beginning and Intermediate Algebra, Building a Foundation* by McKenna/Kirk is represented by at least 4 full pages in this workbook. The first two pages are designated for taking notes, defining key terms and procedures, explaining topics in students' own words, and working through guided examples that correlate to the examples in the text. Each guided example presents all the steps of a solution. Students must complete each example by either entering the appropriate math in the space provided or by filling in the missing annotations to explain in their own words why a particular step is necessary. The remaining pages provide many practice problems to assess students' understanding of the topics at hand. The odd answers to these practice problems are provided at the back of the supplement, and a complete set of answers is available to instructors on-line.

We encourage you to customize your use of these worksheets and use the components whatever way is most beneficial. Students can use these worksheets in the following ways:

- Take classroom notes: write key definitions, terms, and procedures in the space provided.
- Define key terms and procedures in your own words.
- Read the worked out examples in your book or in class, then work through the corresponding guided example for extra practice before starting your homework.
- Write your own notes and words of advice from your instructor next to the key topics.
- Use the additional practice problems as a way to "show your work" or as a quick assessment in class before starting the homework assignment.

In addition to the suggestions above, we hope you will take full advantage of all that this workbook has to offer. We sincerely hope that this workbook will provide instructors and students alike with the support needed for a successful algebra course!

Name: Date:
Instructor: Section:

Chapter 1 REVIEW OF REAL NUMBERS AND PROBLEM SOLVING

1.1 Sets, Symbols, and Real Numbers

KEY VOCABULARY

Term	Definition	Example
Set		
Variable		
Union of sets		
Intersection of sets		

KEY PROPERTIES, PROCEDURES, OR STRATEGIES

Sets of Numbers	Description
Natural numbers	
Whole numbers	
Integers	
Rational numbers	
Irrational numbers	
Real numbers	

Name: Date:
Instructor: Section:

GUIDED EXAMPLE

Using roster notation, list the members from the following set of numbers that are elements of the sets below.

$$\left\{-72,\ -3\frac{2}{3},\ -1.909909990...,\ -\sqrt{2},\ 0,\ 3,\ \pi,\ 4\frac{4}{9},\ 5.27,\ \sqrt{36},\ 88\right\}$$

a) The set of whole numbers **b)** The set of integers
c) The set of natural numbers **d)** The set of rational numbers
e) The set of irrational numbers **f)** The set of real numbers

Solution

a) The set of whole numbers is the counting numbers and 0. Remember, $\sqrt{36}=6$ because $6^2=36$. Members of the set of whole numbers:

b) The set of integers is the whole numbers and the opposite of all the natural numbers. Members of the set of integers:

c) The set of natural numbers is the counting numbers. Members of the set of natural numbers:

d) Try writing each element of the set as a fraction *a* over *b*, such that *a* and *b* are integers and *b* is not equal to 0, or as a decimal that either terminates or repeats. Elements that have decimal representations that are infinite and do not repeat are irrational numbers. If a number is irrational, leave the answer blank.

$-72=$ ______ $-3\frac{2}{3}=$ ______ $-1.909909990...=$ ______ $-\sqrt{2}=$ ______

$0=$ ______ $3=$ ______ $\pi=$ ______ $4\frac{4}{9}=$ ______

$5.27=$ ______ $\sqrt{36}=$ ______ $88=$ ______

Members of the set of rational numbers:

e) Members of the set of irrational numbers:

f) Every number in the set is either rational or irrational, and therefore a real number. Members of the set of real numbers.

Name: Date:
Instructor: Section:

PRACTICE PROBLEMS

Let $A = \{2, 4, 6, 8, \ldots\}$ *and* $B = \{-4, -2, 0, 1, 3, 5, 7\}$. *Determine if the following statements are true or false. If a statement is false, rewrite it so it becomes true.*

1. $3 \in A$ **1.**________________

2. $0 \in B$ **2.**________________

3. $100 \notin A$ **3.**________________

Let $A = \left\{-95, -77, -8.252252225\ldots, -\sqrt{17}, -\pi, 0, 2, \sqrt{7}, \sqrt{9}, \frac{14}{3}, 5, 8.25, 9.\overline{3}\right\}$.

Use roster notation to list the elements of A that belong in each set.

4. The set of whole numbers **4.**________________

5. The set of rational numbers **5.**________________

6. The set of real numbers **6.**________________

7. The set of integers **7.**________________

8. The set of natural numbers **8.**________________

9. The set of irrational numbers **9.**________________

Name: Date:
Instructor: Section:

Let $A = \{-3, -2, -1, 0, 1, 3, 5\}$, $B = \{x | x \in Z \text{ with } x > 3\}$, $C = \{2, 3, 4, 5\}$, *and* $D = \{x | x \in Z \text{ with } x < -1\}$. *Use roster notation to list the elements of each set.*

10. $A \cup C$ **10.**________________

____

11. B **11.**________________

____

12. $B \cap C$ **12.**________________

____

13. $B \cap D$ **13.**________________

____

Using the ordering principle, determine if the following statements are true or false. If a statement is false, rewrite it so it becomes true.

14. $5 > -2$ **14.**________________

15. $-17 > -8$ **15.**________________

Name: Date:
Instructor: Section:

Chapter 1 REVIEW OF REAL NUMBERS AND PROBLEM SOLVING

1.2 Review of Fractions and Decimals

KEY VOCABULARY

Term	Definition	Example
Prime number		
Composite number		
Prime factorization		
Least common multiple (LCM)		
Numerator		
Denominator		
Improper fraction		
Mixed number		
Reciprocals		
Least common denominator (LCD)		

Name: Date:
Instructor: Section:

KEY PROPERTIES, PROCEDURES, OR STRATEGIES

Find the LCM of Two or More Natural Numbers

In the Language of Math	In Your Own Words

Multiply Decimals

In the Language of Math	In Your Own Words

Divide Decimals

In the Language of Math	In Your Own Words

GUIDED EXAMPLE

Subtract $\frac{7}{8}-\frac{1}{5}$.

Solution

LCD (8, 5) = $\boxed{}$ **Find the LCD of 8 and 5.**

$\frac{7}{8}-\frac{1}{5}=\frac{?}{\boxed{}}-\frac{?}{\boxed{}}$ **Write equivalent fractions with the LCD.**

$\boxed{}$ **Multiply by special forms of 1 to obtain the LCD.**

$\boxed{}$ **Write each fraction with a denominator equal to the LCD.**

$\boxed{}$ **Add the numerators and keep the common denominator.**

Write as a reduced fraction if necessary.

Name: Date:
Instructor: Section:

PRACTICE PROBLEMS

1. Use a factor tree to find the prime factorization of 540. **1.**_______________

2. Find LCM (8, 15). **2.**_______________

3. Simplify $\frac{16}{144}$. **3.**_______________

4. Multiply and write the answer as a simplified proper or improper fraction. **4.**_______________

$5\frac{2}{3}\cdot 7\frac{1}{8}$

5. Divide and write the answer as a simplified proper or improper fraction. **5.**_______________

$\frac{3}{8}\div\frac{6}{5}$

6. Add and reduce the answer to lowest terms. **6.**_______________

$\frac{3}{14}+\frac{5}{14}$

7. Subtract and reduce the answer to lowest terms. **7.**_______________

$\frac{18}{25}-\frac{3}{25}$

8. Find the LCD. Do not add the fractions. **8.**_______________

$\frac{4}{5}+\frac{7}{12}$

Name: Date:
Instructor: Section:

Add or subtract, and reduce the answers to lowest terms. Write all answers as proper or improper fractions.

9. $\frac{7}{8}+\frac{4}{9}$ **9.**________________

10. $\frac{5}{14}-\frac{1}{8}$ **10.**________________

11. $8\frac{1}{9}-5\frac{1}{6}$ **11.**________________

Perform the indicated operations.

12. $54.025+16.76$ **12.**________________

13. $77.21-58.496$ **13.**________________

14. $(8.9)(4.56)$ **14.**________________

15. $32.032\div18.2$ **15.**________________

Name: Date:
Instructor: Section:

Chapter 1 REVIEW OF REAL NUMBERS AND PROBLEM SOLVING

1.3 Addition and Subtraction of Real Numbers

KEY VOCABULARY

Term	Definition	Example
Absolute value		
Additive inverse		

KEY PROPERTIES, PROCEDURES, OR STRATEGIES

Commutative Property of Addition

In the Language of Math	In Your Own Words

Rules for Adding Real Numbers

In the Language of Math	Examples

Rules for Subtracting Real Numbers

In the Language of Math	Examples

Name: Date:
Instructor: Section:

GUIDED EXAMPLES

1. Add the following numerals using two methods: adding left to right and using the Commutative Property of Addition.

$(-10)+6+14+(-17)+(-5)$

Solution

Left to right:

$(-10)+6+14+(-17)+(-5)$

$= \boxed{} +14+(-17)+(-5)$

$= \boxed{} +(-17)+(-5)$

$= \boxed{} +(-5)$

$= \boxed{}$

Using the Commutative Property of Addition:

$(-10)+6+14+(-17)+(-5)$

$= 6 + 14 + \boxed{}$

$= \boxed{} + \boxed{}$

$= \boxed{}$

2. Perform the indicated operations.

$15+(-6)-7-(-18)+4+(-9)$

Solution

$15+(-6)-7-(-18)+4+(-9)$ — **Identify the operations of subtraction.**

$= \boxed{}$ — **Rewrite as addition of the opposite.**

$= \boxed{}$ — **Apply the Commutative Property of Addition.**

$= \boxed{}$ — **Add all the positive numbers, then add all the negative numbers.**

$= \boxed{}$ — **Subtract the numbers and attach the sign of the larger absolute value.**

NOTES

Name: Date:
Instructor: Section:

PRACTICE PROBLEMS

Determine the absolute value.

1. $|-17|$ **1.**________________

2. $|53|$ **2.**________________

Perform the indicated operations.

3. $45+(-15)$ **3.**________________

4. $-27+(-18)$ **4.**________________

5. $-90+66$ **5.**________________

6. $-15.6+(-14.3)$ **6.**________________

7. $-\frac{1}{2}+\left(-\frac{3}{4}\right)$ **7.**________________

8. $-72-85$ **8.**________________

9. $14-(-17)$ **9.**________________

Name: Date:
Instructor: Section:

10. $-11.7-(-3.6)$ **10.**________________

11. $\frac{3}{5}-\left(-\frac{7}{8}\right)$ **11.**________________

12. $12-(-13)+25-17-(-18)$ **12.**________________

13. $-78-(-80)+12-15+(-16)$ **13.**________________

14. $\left(-\frac{16}{30}\right)-(-2)$ **14.**________________

15. $-226.5+(-142.8)+97$ **15.**________________

Name: Date:
Instructor: Section:

Chapter 1 REVIEW OF REAL NUMBERS AND PROBLEM SOLVING

1.4 Multiplication, Division, and Properties of Real Numbers

KEY PROPERTIES, PROCEDURES, OR STRATEGIES

Commutative Property of Multiplication

In the Language of Math	In Your Own Words

Associative Property of Addition

In the Language of Math	In Your Own Words

Associative Property of Multiplication

In the Language of Math	In Your Own Words

Distributive Property

In the Language of Math	In Your Own Words

Identity Property of Addition

In the Language of Math	In Your Own Words

Identity Property of Multiplication

In the Language of Math	In Your Own Words

Inverse Property of Addition

In the Language of Math	In Your Own Words

Inverse Property of Multiplication

In the Language of Math	In Your Own Words

Name: Date:
Instructor: Section:

GUIDED EXAMPLES

In the table below, complete the example with the appropriate algebraic expression, identify the illustrated property, or justify why the property holds for the given example.

Example	**Property**	**Justification**
$(-5)7 = 7(-5)$		Order changed
$(x+4)+7 =$ ________	Associative Property of Addition	Grouping changed
$8+(-8)=0$	Inverse Property of Addition	
$5(a+b) =$ ________	Distributive Property	Distribute the 5
$9\left(\frac{1}{9}\right)=1$	Inverse Property of Multiplication	
$7(4x) =$ ________	Associative Property of Multiplication	Grouping changed
$(7+0)+2=7+2$		$7+0=7$
$-6+8=8+(-6)$	Commutative Property of Addition	
$(2\cdot 1)(-7)=2(-7)$		$2\cdot 1=2$

NOTES

Name: Date:
Instructor: Section:

PRACTICE PROBLEMS

Apply the given property of real numbers to fill in the blank with the appropriate algebraic expression.

1. Identity Property of Multiplication **1.**__________

$\left(\frac{1}{12}\right)\cdot 1 =$ __________

2. Commutative Property of Addition **2.**__________

$7a+8d=$ __________

3. Distributive Property **3.**__________

$12(c+d)=$ __________

4. Associative Property of Multiplication **4.**__________

$(4\cdot 5)6=$ __________

5. Identity Property of Addition **5.**__________

$\left(\frac{1}{5}+0\right)+\frac{1}{4}=$ __________

6. Inverse Property of Multiplication **6.**__________

$\frac{1}{5}\cdot$ __________ $=1$

7. Associative Property of Addition **7.**__________

$7p+(5q+3r)=$ __________

8. Inverse Property of Addition **8.**__________

$12+$ __________ $=0$

9. Commutative Property of Multiplication **9.**__________

$8a+(4+7)=$ __________

Name: Date:
Instructor: Section:

Perform the indicated operations.

10. $(-17)\cdot(15)$

10.________________

11. $0(45)$

11.________________

12. $\frac{144}{-6}$

12.________________

13. $(-185.84)\div(-10.1)$

13.________________

14. $-7\left(-\frac{3}{4}\right)$

14.________________

15. $16\div\left(-\frac{8}{5}\right)$

15.________________

Name: Date:
Instructor: Section:

Chapter 1 REVIEW OF REAL NUMBERS AND PROBLEM SOLVING

1.5 Exponents, Order of Operations, and Algebraic Expressions

KEY VOCABULARY

Term	Definition	Example
Exponential expression		
Algebraic expression		
Term		
Coefficient		
Constant term		
Like terms		

KEY PROPERTIES, PROCEDURES, OR STRATEGIES

Order of Operations

1.

2.

3.

4.

Name: Date:
Instructor: Section:

GUIDED EXAMPLES

1. Simplify: $\dfrac{4^2-5(-3+8)}{\sqrt{7-3}}$

Solution

$\dfrac{4^2-5(-3+8)}{\sqrt{7-3}}$

$=\dfrac{4^2-5(5)}{\sqrt{4}}$ [ ]

$=\dfrac{4^2-5(5)}{2}$ **Evaluate the square root.**

$=\dfrac{16-5(5)}{2}$ [ ]

$=\dfrac{16-25}{2}$ [ ]

$=\dfrac{-9}{2}$ **Perform subtraction in the numerator.**

$=-4.5$ **Divide.**

2. Evaluate $8x^2-5y+3z-10$ when $x=-4, y=3, z=8$.

Solution

$8x^2-5y+3z-10$

$=8(\ )^2-5(\ )+3(\ )-10$ **Replace the variables with parentheses.**

$=$ [ ] **Substitute the values for each variable.**

$=$ [ ] **Evaluate the exponent.**

$=$ [ ] **Multiply.**

$=$ [ ] **Add and subtract, left to right.**

$=$ [ ] **Simplify.**

Name: Date:
Instructor: Section:

Chapter 1 REVIEW OF REAL NUMBERS AND PROBLEM SOLVING

1.6 Reading and Interpreting Graphs and the Coordinate Plane

KEY VOCABULARY

Term	Definition	Example
Graph		
Pictograph		
Bar graph		
Circle graph		
Line graph		
Cartesian Coordinate Plane		
Origin		

Name: Date:
Instructor: Section:

GUIDED EXAMPLES

1. This circle graph shows the number of different colored jelly beans in one package. Use the circle graph to answer the questions that follow.
   a. What percent of jelly beans are purple?
   b. How many jelly beans in the package were pink?

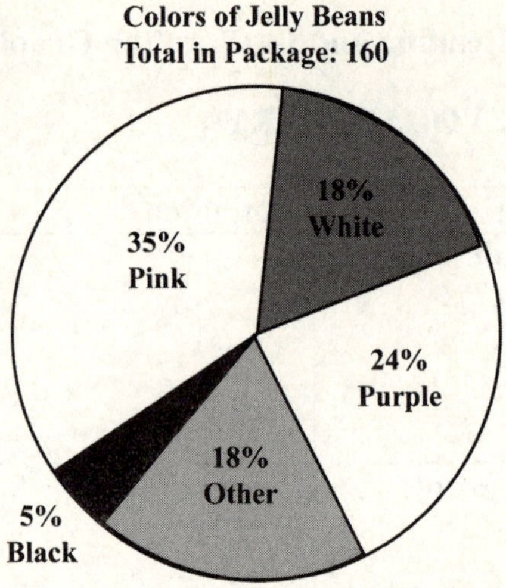

Solution

a. The portion of the graph that measures 24% is designated as purple.

b. The pink portion is 35% of the total number of jelly beans in the bag. The word "of" means to ____________________, so the equation we use to find the number of pink jelly beans is [ ]

The number of pink jelly beans is [ ]

2. This line graph shows the number of fruit baskets sold by Red Bird Troop 612 over several years. Use the line graph to answer the questions that follow.

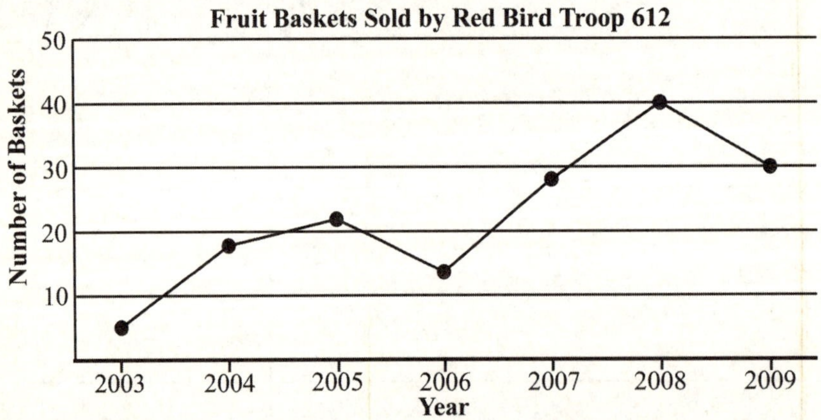

a. In what years did the troop sell more than 25 fruit baskets?

b. How many fewer baskets did the troop sell in 2009 than in 2008?

Solution

a. Use a straightedge to draw a horizontal line from the value 25 across the graph. The dots that lie above this line represent the years the troop sold more than 25 fruit baskets. These years are [ ]

b. In 2009 the troop sold [ ] baskets and the troop sold [ ] baskets in 2008. The difference between these two values is [ ]

Name: Date:
Instructor: Section:

PRACTICE PROBLEMS

Refer to the pictograph representing the number of books read by Mr. Sutton's eighth grade class.

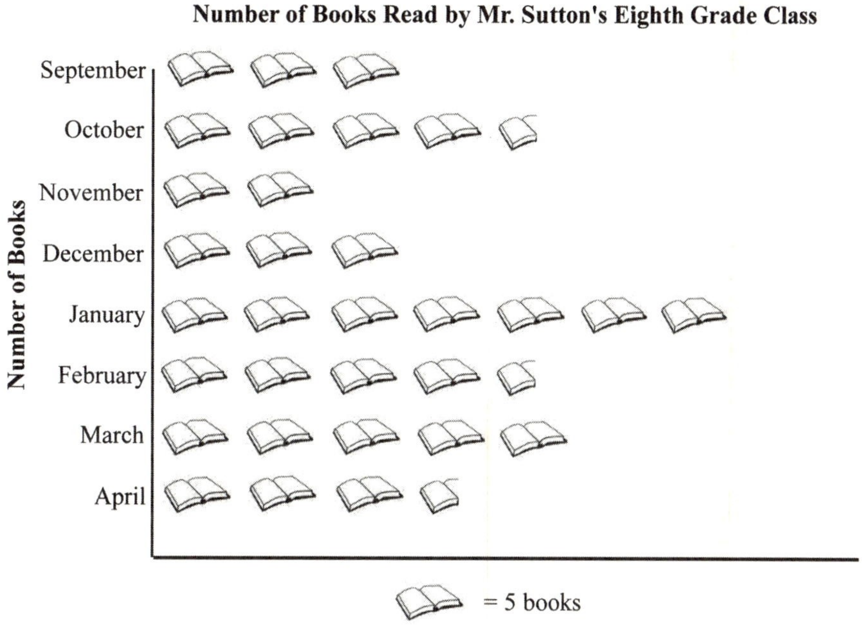

1. How many books did the class read in January? **1.**____________

2. How many more books did the class read in March than in November? **2.**____________

Refer to the bar graph representing the number of people that visited the website for the East Seneca Central Schools during one week.

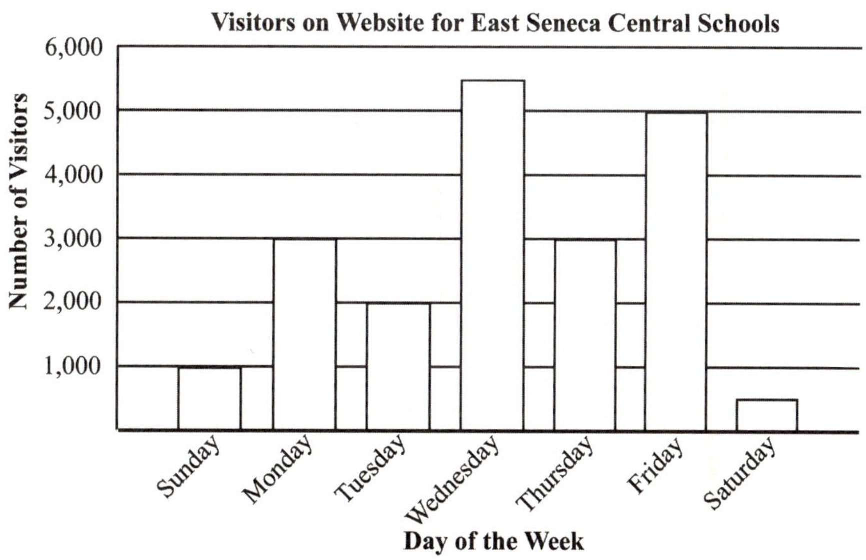

Name: Date:
Instructor: Section:

3. Approximately how many people visited the website on Monday? **3.**______________

4. How many more people visited the website on Friday than on Tuesday? **4.**______________

5. On Saturday, 30 people visited the website from the town library. What percent of the people that visited the website on Saturday did so at the library? **5.**______________

Refer to the circle graph representing the Student Government Association's yearly budgeted expenses.

Student Government Association Yearly Budget
Total Budget $112,500

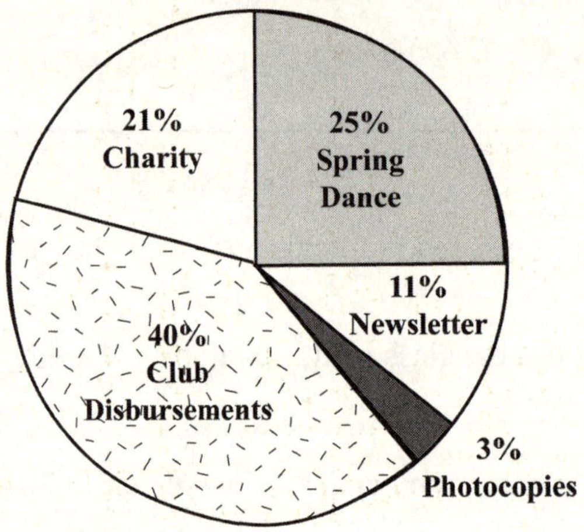

6. What percent of the budget is set aside for photocopies? **6.**______________

7. How much money is set aside for club disbursements? **7.**______________

8. Plot the points on the coordinate plane. Label each point clearly and state in which quadrant it lies (if any). **8.**

a. $(-4,2)$

b. $(0,-6)$

c. $(2,8)$

d. $(4,-4)$

e. $(-3,-6)$

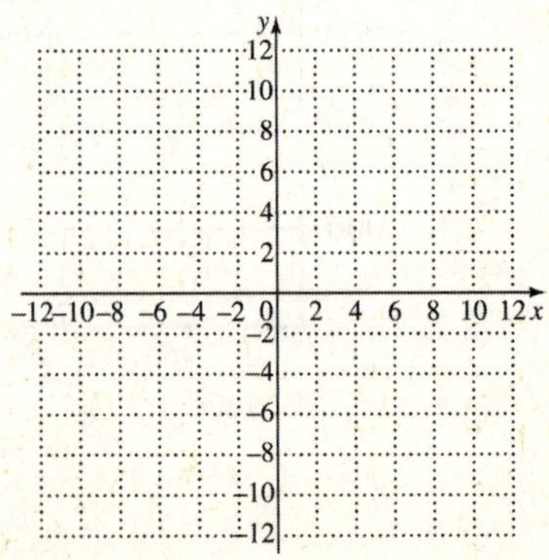

Name: Date:
Instructor: Section:

Chapter 1 REVIEW OF REAL NUMBERS AND PROBLEM SOLVING

1.7 Problem Solving: The 4P's Strategy

KEY PROPERTIES, PROCEDURES, OR STRATEGIES

The 4P's Problem Solving Strategy

Step 1:	In Your Own Words
Step 2:	In Your Own Words
Step 3:	In Your Own Words
Step 4:	In Your Own Words

Name: Date:
Instructor: Section:

GUIDED EXAMPLE

Find the next three apparent terms in the sequence 36, 108, 324, 972, ….

Solution

1 Prepare

We are asked to determine

2 Plan

Use the *look for a pattern* strategy.

3 Process

Study the given numbers. Do the numbers increase or decrease?

What operation(s) can we use to create a pattern of increasing/decreasing numbers?

Compare the first two numbers.

What is their difference?

What is their quotient?

Compare the second number and the third number.

What is their difference?

What is their quotient?

Which of these values is the same as for the first two numbers?

Do you have an idea of what the pattern might be?

What are the next three numbers in the sequence?

4 Ponder

Does the answer seem reasonable? How can you verify that the pattern you found is correct?

Name: Date:
Instructor: Section:

PRACTICE PROBLEMS

Solve each problem by utilizing the 4P's problem-solving strategy.

1. Meghan has 11 coins in her car's console made up of nickels, dimes, and quarters. If the total value of the coins is $1.85, how many of each coin does she have?

1.________________

2. Find two positive numbers whose sum is 118 and whose product is 3,477.

2.________________

3. Find the next three apparent terms in the sequence 59, 66, 73, 80, 87, 94, ...

3.________________

4. Find the next three apparent terms in the sequence 625, 125, 25, 5, ...

4.________________

5. Zachary is getting ready to plant vegetables in his garden. He wants to plant twice as many carrots as lettuce plants and two fewer tomato plants than the number of carrot plants. If he purchases 18 plants, how many of each type will there be?

5.________________

Name: Date:
Instructor: Section:

6. A discount shoe store is setting up a display of shoe boxes. There are 8 layers and each layer has one less box than the layer below it. If the bottom layer has 10 boxes, how many total boxes are there?

6.________________

7. Amanda, Greenlee, Kendall, and Annie are modeling at a fashion event. Amanda walked down the runway ahead of Kendall but behind Annie. Greenlee was the third person down the runway. Who walked second?

7.________________

8. The Little Folks preschool is having a parade in which the children will ride tricycles and bicycles with training wheels. There are 15 children in the parade. If you count 49 wheels altogether, how many of each vehicle are in the parade?

8.________________

9. The library tracks the number of books borrowed each week by genre. Make a bar graph that shows the following data in a meaningful way.

9.________________

Genre	Number of Books Borrowed
Mystery	67
Travel	14
Romance	42
Biography	22
History	30

Name: Date:
Instructor: Section:

Chapter 2 EQUATIONS AND INEQUALITIES IN ONE VARIABLE WITH APPLICATIONS

2.1 Properties of Equality and Linear Equations

KEY VOCABULARY

Term	Definition	Example
Equation		
Solution of an equation		
Conditional linear equation in one variable		

KEY PROPERTIES, PROCEDURES, OR STRATEGIES

Symmetric Property of Equality

In the Language of Math	In Your Own Words

Addition Property of Equality

In the Language of Math	In Your Own Words

Multiplication Property of Equality

In the Language of Math	In Your Own Words

Name: Date:
Instructor: Section:

Strategy for Solving Linear Equations

Procedure

1.

2.

3.

4.

5.

6.

GUIDED EXAMPLES

Solve the following linear equations for the indicated variable and check the solutions.

a) $3x+5=11$

Solution

$$3x+5=11$$

***Isolate the variable quantity*, 3*x*, on the LHS.**

$$3x+5-5=11-5$$

$$3x=6$$

$$\frac{{}^{1}\not{3}x}{\not{3}_{1}}=\frac{6}{3}$$

***Solve* for the variable by dividing each side by 3.**

$$x=2$$

Achieve goal: $1\cdot x=?$

Solution set: $\{2\}$

Check: $3x+5=11$

$$3(2)+5\stackrel{?}{=}11$$

$$11=11$$

b) $-7x-12=8x+3$

Solution

$$-7x-12=8x+3$$

***Gather the variables* to the RHS.**

Add 7*x* to both sides to get a positive coefficient for *x* on the RHS.

***Isolate* the variable on the RHS.**

***Solve* for the variable by dividing each side by 15.**

Achieve goal: $1\cdot x=?$

Solution set:

Check:

Name: Date:
Instructor: Section:

Chapter 2 EQUATIONS AND INEQUALITIES IN ONE VARIABLE WITH APPLICATIONS

2.2 Linear Equations Involving Grouping Symbols and Fractions

KEY VOCABULARY

Term	Definition	Example
Inconsistent equation		
Identity		

REVIEW KEY PROPERTIES, PROCEDURES, OR STRATEGIES

Distributive Property

In the Language of Math	In Your Own Words

Order of Operations

In the Language of Math	In Your Own Words

Name: Date:
Instructor: Section:

GUIDED EXAMPLES

1. Solve the following linear equation and check the solution.
$5x+0.7(2x-0.1)=127.93$

Solution

$5x+0.7(2x-0.1)=127.93$	***Eliminate* parentheses using the Distributive Property.**
$5x+1.4x-0.07=127.93$	
$6.4x-0.07=127.93$	**Combine like terms.**
$6.4x-0.07+0.07=127.93+0.07$	
$6.4x=128$	
$10(6.4x)=10(128)$	
$64x=1280$	
$\frac{\cancel{64}x}{\cancel{64}}=\frac{1280}{64}$	
$x=20$	**Achieve goal:** $1\cdot x=?$
Solution set: $\{20\}$	**Check:**

2. Set up an algebraic equation and solve it.
Two angles are supplementary if the sum of their measures is $180°$. If two angles are supplementary and the measure of one angle is four more than three times the measure of the other angle, find the two angles.

Solution

Let x = measure of angle 1. Then measure of angle 2 =

State the equation in words.

Write the algebraic equation.

***Isolate* the variable on one side.**

***Solve* for the variable.**

Achieve goal: $1\cdot x=?$

Measure of angle 1 =

Measure of angle 2 =

Check:

Name: Date:
Instructor: Section:

Chapter 2 EQUATIONS AND INEQUALITIES IN ONE VARIABLE WITH APPLICATIONS

2.3 Literal Equations and Applications

KEY VOCABULARY

Term	Definition	Example
Literal equation		
Formula		

KEY PROPERTIES, PROCEDURES, OR STRATEGIES

Simple Interest Formula

In the Language of Math	In Your Own Words

Distance-Rate-Time Formula

In the Language of Math	In Your Own Words

NOTES

Name: Date:
Instructor: Section:

GUIDED EXAMPLES

1. At what annual rate must \$8,520 be invested to earn \$798.75 in interest in 15 months?
Solution

$I =$

$P =$

$r =$

$t =$

Solve $I = Prt$ for ________

***Isolate* the unknown variable.**

***Substitute* the known values.**

***Solve* for the unknown value.**

Annual rate:

2. How fast is a commercial jet flying, if it travels 1050 miles in 2 hours 30 minutes?
Solution

$D =$

$R =$

$T =$

***State* the formula.**

***Substitute* the known values.**

***Solve* for the unknown value.**

Average speed:

Name: Date:
Instructor: Section:

Chapter 2 EQUATIONS AND INEQUALITIES IN ONE VARIABLE WITH APPLICATIONS

2.4 Linear Inequalities

KEY VOCABULARY

Term	Definition	Example
Linear inequality in one variable		
Solution set for a linear inequality		
Inconsistent inequality		
Identity inequality		

KEY PROPERTIES, PROCEDURES, OR STRATEGIES

Addition Property of Inequalities

In the Language of Math	In Your Own Words

Multiplication Property of Inequalities

In the Language of Math	In Your Own Words

Name: Date:
Instructor: Section:

GUIDED EXAMPLES

1. Solve the following linear inequality for the indicated variable and graph the solution set on the real number line. Express the solution set as an inequality involving x and in interval notation.
$-2(x+3)<4(3x-5)$

Solution
$-2(x+3)<4(3x-5)$

[ ]	**Simplify by distributing on both sides.**
[ ]	**Gather variables to the LHS.**
[ ]	**Isolate the variable.**
[ ]	**Solve for the variable.** **Achieve goal: x is on the LHS with a coefficient of positive 1.**

Graph: ⟷

Interval Notation: [ ]

2. Solve the following linear inequalities and state the solutions sets as intervals, if applicable.
a) $2x-7>2x$ **b)** $-x+4>2-x$

Solution

a) $2x-7>2x$

[ ] **Gather variables to the LHS.**

This is an ________________ inequality.
The solution set is [ ]

b) $-x+4>2-x$

[ ] **Gather variables to the LHS.**

This is an ________________ inequality.
The solution set is [ ]

Name: Date:
Instructor: Section:

Chapter 2 EQUATIONS AND INEQUALITIES IN ONE VARIABLE WITH APPLICATIONS

2.5 Compound and Double Inequalities

KEY VOCABULARY

Term	Definition	Example
Compound inequality		
Double inequality		

NOTES

Name: Date:
Instructor: Section:

GUIDED EXAMPLES

1. Solve the following compound inequality by graphing the solution on the real line. Express the answer using interval notation.
$x-5<7$ or $-2x>8$
Solution
Solve each inequality for *x*.

Graph each inequality on a single real line. You may use two different colors if it is helpful.

Consider what the connective "or" means. In interval notation, the solution is

2. Solve the following double inequality and graph the solution set on the real line. State the answer using interval notation.
$7<6-8x<15$
Solution
$7<6-8x<15$

Isolate the variable quantity in the middle by subtracting _____ from all three parts.

Solve for the variable quantity with a coefficient of positive 1. Divide all three parts by _____ and ___________ both of the inequality symbols.

Write the double inequality from least to greatest.

Graph:

Interval Notation:

Name: Date:
Instructor: Section:

Chapter 2 EQUATIONS AND INEQUALITIES IN ONE VARIABLE WITH APPLICATIONS

2.6 Absolute Value Equations and Inequalities

KEY VOCABULARY

Term	Definition	Example
Absolute value of the real number *x*		

KEY PROPERTIES, PROCEDURES, OR STRATEGIES

General Form of an Absolute Value Equation

In the Language of Math	In Your Own Words

General Form for an Absolute Value Involving "Less Than"

In the Language of Math	In Your Own Words

General Form for an Absolute Value Involving "Greater Than"

In the Language of Math	In Your Own Words

Name: Date:
Instructor: Section:

GUIDED EXAMPLES

1. Solve the following absolute value equation.

$|8x-5|+7=19$

Solution

$|8x-5|+7=19$

Isolate the absolute value on the LHS.

or

Apply the general form to get two linear equations and solve for *x*.

Solution set:

2. Solve the following absolute value inequality. Graph the solution set on the real number line and state the solution using interval notation.

$|-2x+5|\geq 13$

Solution

$|-2x+5|\geq 13$

or

Set up the compound inequality using the connective *or*.

Solve each inequality.

Graph:

Interval Notation:

Name: Date:
Instructor: Section:

Chapter 3 LINEAR EQUATIONS, SLOPE, INEQUALITIES, AND INTRODUCTION TO FUNCTIONS

3.1 Linear Equations in Two Variables

KEY VOCABULARY

Term	Definition	Example
Linear equation in two variables		
Solution to a linear equation		
Graph of a linear equation in two variables		

KEY PROPERTIES, PROCEDURES, OR STRATEGIES

Graphing Method 1: The Plotting Points Method

	Example
1. Make a T table with three or more solutions. **2.** Graph the ordered pairs on the plane. **3.** Connect the solutions using a straightedge	

Name: Date:
Instructor: Section:

GUIDED EXAMPLES

1. Determine if the ordered pair $(8,-2)$ is a solution to the linear equation $4x+5y=22$.

Solution

$4x+5y=22$	**Rewrite the original equation.**
$4(\quad)+5(\quad)\overset{?}{=}22$	**Substitute the values for *x* and *y*.**
$\square=22$	**Simplify the LHS.**
$\square=22$	**Is the statement true or false?** $\square$

2. Graph the solutions of the linear equation $2x+3y=3$ by filling in the T table.

x	***y***
0	?
?	0
−3	?

Solution

Substitute $x=0$ and solve for *y*.	Substitute $y=0$ and solve for *x*	Substitute $\square$ and solve for $\square$
$2x+3y=3$ $2(\quad)+3y=3$ $\square=\square$ $\square=\square$	$2x+3y=3$ $2x+3(\quad)=3$ $\square=\square$ $\square=\square$	

Now complete the T table:

x	*y*
0	
	0
−3	

Plot the three points and connect using a straightedge.

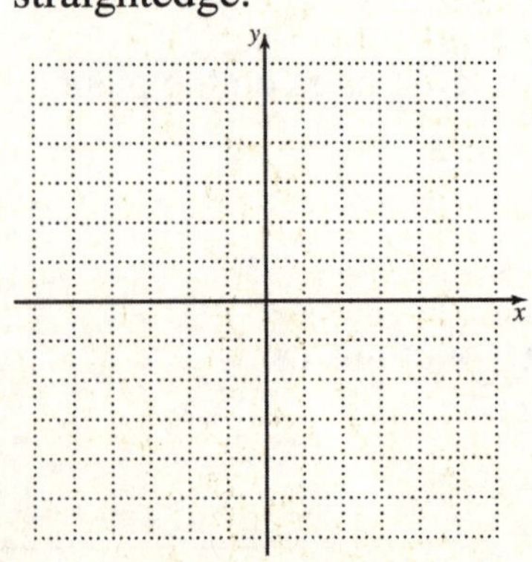

Name: Date:
Instructor: Section:

PRACTICE PROBLEMS

Determine which of the following represent a linear equation in two variables. If an equation is linear, rewrite it in standard form Ax + By = C. If an equation is not linear, explain why.

1. $x=\frac{1}{5}y-4$ **1.**______________

2. $3xy+2y=7$ **2.**______________

Determine if each ordered pair is a solution to the given equation.

3. $(5,-1)$ $-2x+4y=-14$ **3.**______________

4. $(7,1)$ $3x=y+14$ **4.**______________

5. $\left(\frac{11}{2},4\right)$ $5y=4x-2$ **5.**______________

Find the missing coordinate of the solutions for the given equation.

6. $5x-12y=13$ $(?,1)$ **6.**______________

7. $7x=3y+6$ $(4,?)$ **7.**______________

8. $y=\frac{1}{3}x-4$ $(9,?)$ **8.**______________

9. $0.4x+y=8$ $(?,5)$ **9.**______________

Name: Date:
Instructor: Section:

Graph the solutions to the linear equations by plotting points.

10. $4x+6y=12$

10.

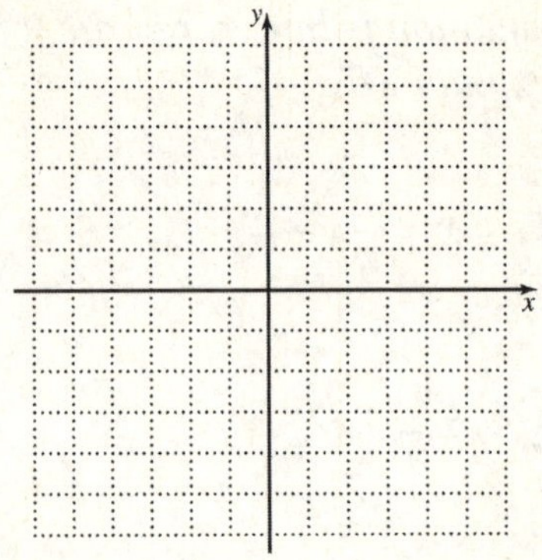

11. $y=2x-5$

11.

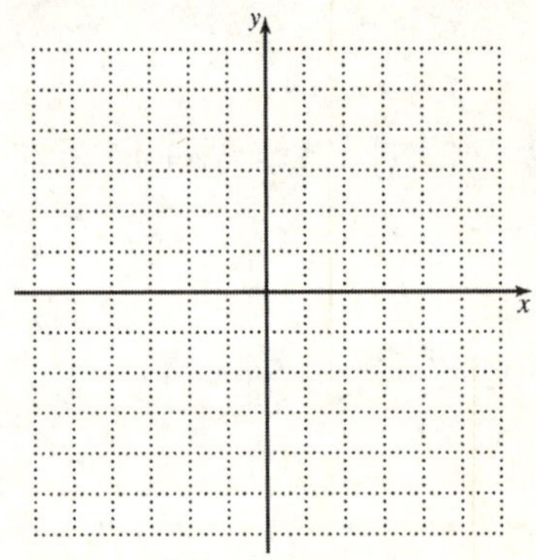

12. $\frac{1}{5}x+y=-2$

12.

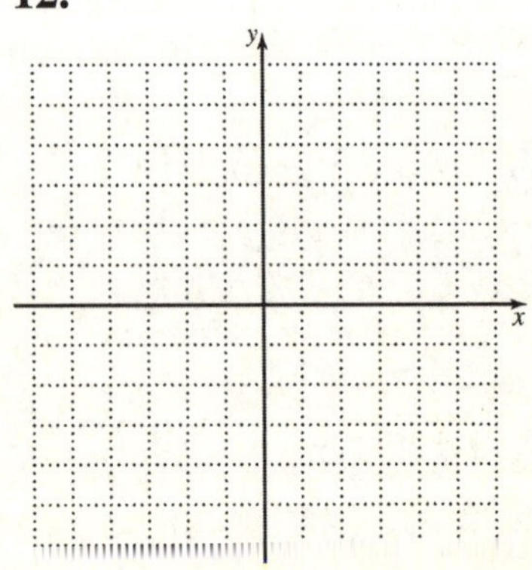

Name: Date:
Instructor: Section:

Chapter 3 LINEAR EQUATIONS, SLOPE, INEQUALITIES, AND INTRODUCTION TO FUNCTIONS

3.2 The Intercept Method

KEY VOCABULARY

Term	Definition	Example
x-intercept		
y-intercept		
Vertical line		
Horizontal line		

KEY PROPERTIES, PROCEDURES, OR STRATEGIES

Graphing Method 2: The Intercept Method

	Example
1. Find the x- and y-intercepts by making a T table. **2.** If the x- and y-intercepts are unique, then graph them both. If they are not unique, find an additional solution and graph it. **3.** Connect the solutions using a straightedge.	

Name: Date:
Instructor: Section:

GUIDED EXAMPLE

Use the intercept method to graph the solutions of the linear equation $4x-3y=12$.

Solution

Find the x- and y-intercepts by filling in the T table.

	x	y	**Solution**
y-intercept	0	?	(0, ?)
x-intercept	?	0	(?, 0)

Set $x = 0$ and solve for y	Set $y = 0$ and solve for x
$4x-3y=12$	$4x-3y=12$

Plot the two points and connect with a straight line.

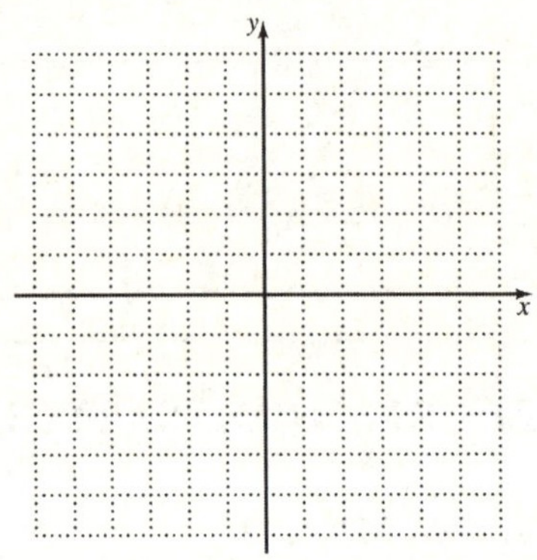

NOTES

Name: Date:
Instructor: Section:

PRACTICE PROBLEMS

Use the intercept method to graph the solutions of the linear equations.

1. $-4x+2y=-12$

1.

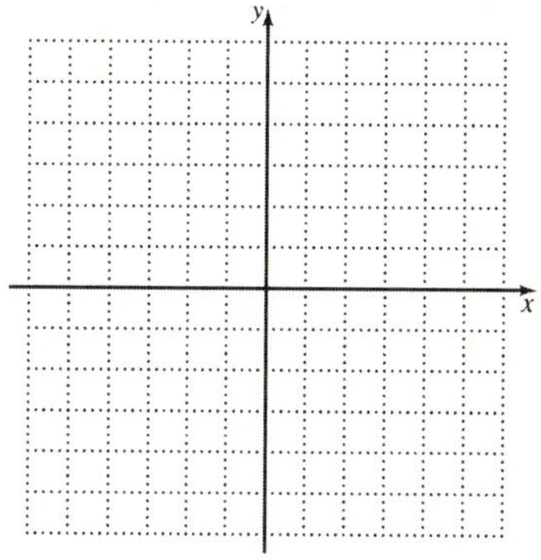

2. $y=-4x-2$

2.

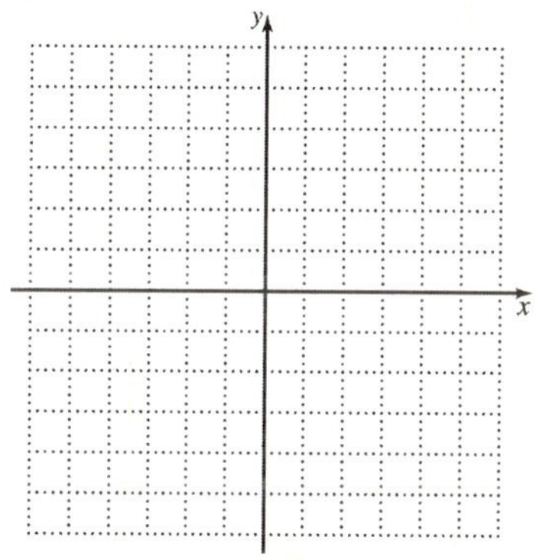

3. $5y=3x-30$

3.

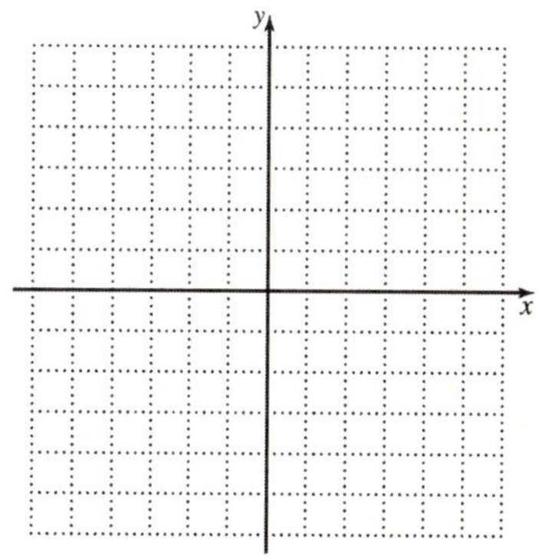

Name: Date:
Instructor: Section:

4. $y = -x$

4.

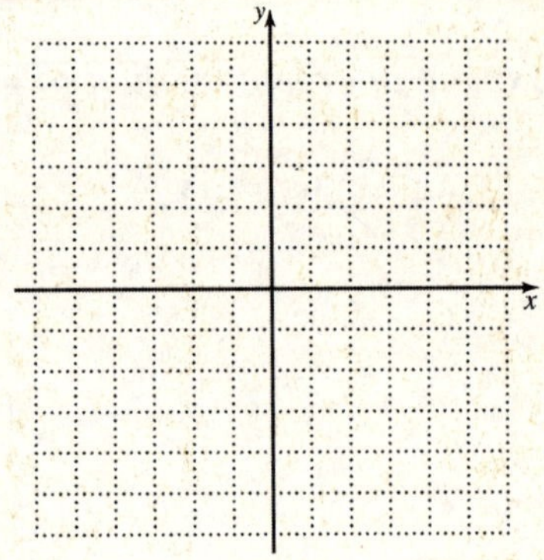

5. $1.5x - 2y = -6$

5.

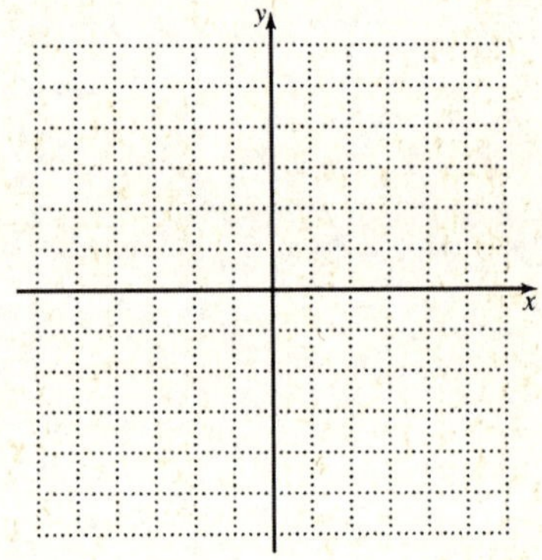

6. $y - 4 = 0$

6.

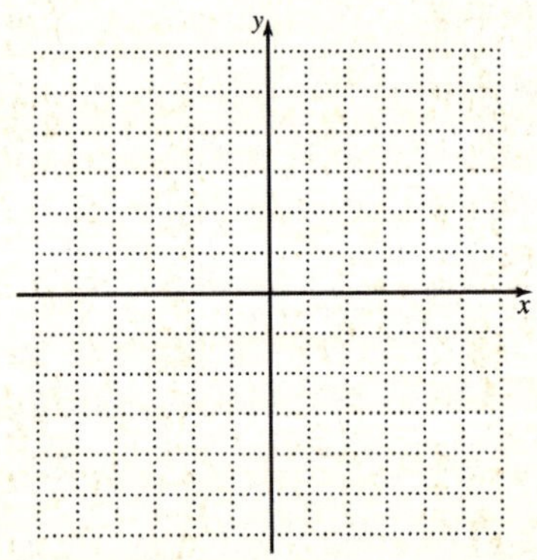

Determine if each graph will be a vertical line, horizontal line, or neither.

7. $x = -15$ **7.**________________

8. $y + 7 = -2$ **8.**________________

9. $-2x + 3 = 1$ **9.**________________

10. $y = -4x$ **10.**________________

Name: Date:
Instructor: Section:

Chapter 3 LINEAR EQUATIONS, SLOPE, INEQUALITIES, AND INTRODUCTION TO FUNCTIONS

3.3 The Slope and *y*-Intercept Method

KEY VOCABULARY

Term	Definition	Example
Slope of a straight line		
Slope-intercept form		

KEY PROPERTIES, PROCEDURES, OR STRATEGIES

Slopes of Horizontal and Vertical Lines

In the Language of Math	In Your Own Words

Graphing Method 3: The Slope and *y*-Intercept Method

	Example
1. Write the equation in $y = mx + b$ form by solving for *y*, if necessary. **2.** Determine the slope and *y*-intercept of the line. **3.** Plot the *y*-intercept. **4.** From the *y*-intercept, graph additional solutions by using $m = \frac{\text{Rise}}{\text{Run}}$.	

Name: Date:
Instructor: Section:

GUIDED EXAMPLES

1. Find the slope of the straight line between the two solutions $(3,-1)$ and $(8,-5)$ using the formula $m=\dfrac{y_2-y_1}{x_2-x_1}$.

Solution

We have two solutions, $(3,-1)$ and $(8,-5)$.

Label them as follows: $\underset{x_1}{\uparrow}\ \underset{y_1}{\uparrow}$ $\underset{x_2}{\uparrow}\ \underset{y_2}{\uparrow}$

Now use the formula.

$m=\dfrac{y_2-y_1}{x_2-x_1}$. **Write the formula for slope.**

$m =$ ☐ **Substitute the values for each solution.**

$m =$ ☐ **Simplify the numerator and denominator.**

$m =$ ☐ **Write as a reduced fraction, if necessary.**

2. Graph the solutions of the linear equation by using the slope and y-Intercept Method.
$3y=-4x+3$

Solution

Start by putting the equation in slope-intercept form, $y=mx+b$, by solving for y.

$3y=-4x+3$

☐ **Divide both sides by** ☐

☐ **Write in slope-intercept form.**

$m =$ ☐

and y-intercept ☐

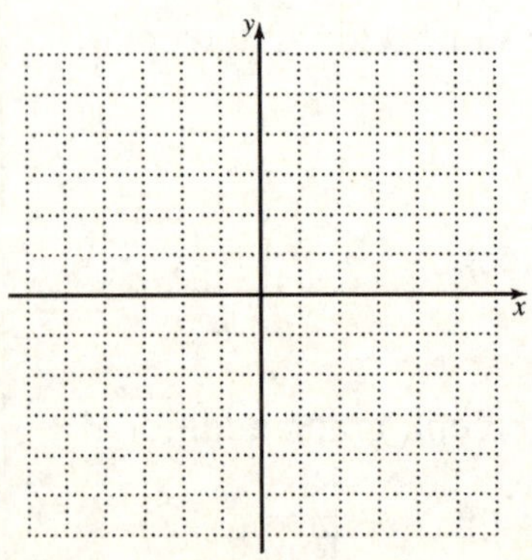

Name: Date:
Instructor: Section:

PRACTICE PROBLEMS

Determine the slope of each line using $m = \dfrac{\text{Rise}}{\text{Run}}$.

1. **1.**________________

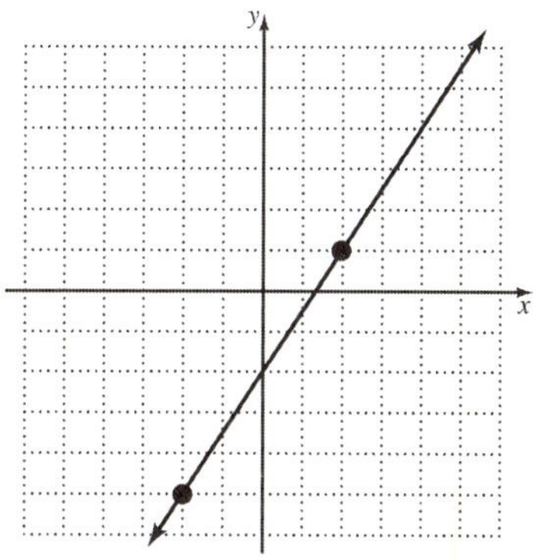

2. **2.**________________

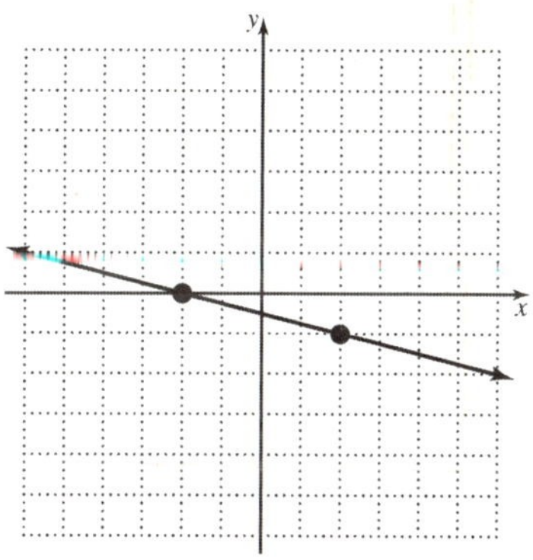

Find the slope of the straight line between the two solution points.

3. $(1,-4)$ and $(9,2)$ **3.**________________

4. $(-5,0)$ and $(8,-2)$ **4.**________________

5. $\left(-\dfrac{1}{2},\dfrac{1}{4}\right)$ and $\left(4,-\dfrac{1}{8}\right)$ **5.**________________

6. $(8.3,-4,1)$ and $(-4.2,5.8)$ **6.**________________

Name: Date:
Instructor: Section:

Find the slope and y-intercept by using the slope-intercept form of the equation of each line. If necessary, solve for y first.

7. $y=-7x+1$ **7.**________________

8. $y=\frac{5}{6}x-8$ **8.**________________

9. $y=-\frac{4}{5}x$ **9.**________________

10. $5x-7y=-11$ **10.**________________

11. $-9x+6=0$ **11.**________________

Graph the solutions of the linear equations by using the slope and y-intercept.

12. $y=-3x+4$ **12.**

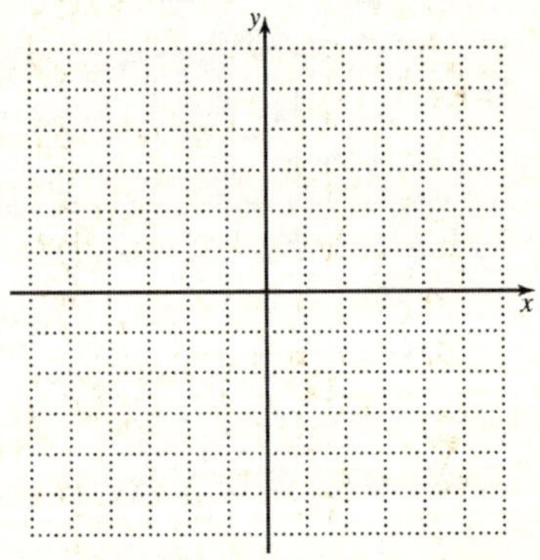

13. $-2x+4y=0$ **13.**

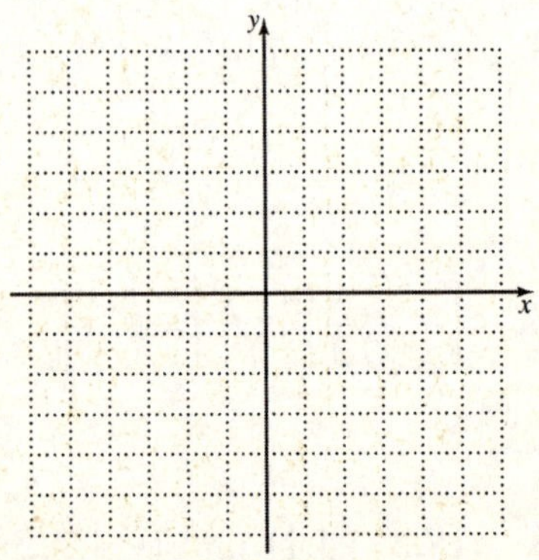

Name: Date:
Instructor: Section:

Chapter 3 LINEAR EQUATIONS, SLOPE, INEQUALITIES, AND INTRODUCTION TO FUNCTIONS

3.4 Determining the Equation of a Line, Parallel Lines, and Perpendicular Lines

KEY VOCABULARY

Term	Definition	Example
Point-slope form		
Parallel lines		
Perpendicular lines		

NOTES

Name: Date:
Instructor: Section:

GUIDED EXAMPLES

1. Write the equation of the line with a slope of $\frac{1}{4}$ that passes through the point $(8,-7)$.

Solution

Find the y-intercept by substituting the known information into the slope-intercept form of the equation of a line.

$y = mx + b$ **Start with the slope-intercept form of the equation.**

$(\quad) = (\quad)(\quad) + (\quad)$ **Substitute the known values in the correct places. Leave the unknown values in variable form.**

Solve the equation to find the value of the y-intercept.

Now that we know the slope and y-intercept, substitute them into the slope-intercept form of the equation, $y = mx + b$.

$y = mx + b$ **Write the slope-intercept form.**

$y = (\quad)x + (\quad)$ **Substitute the slope and y-intercept.**

Write the equation of the line.

2. Determine if the pair of lines are parallel, perpendicular, or neither.

$L_1: -8y = x + 32$

$L_2: x + 8y = 24$

Solution

Begin by writing each equation in slope-intercept form to determine its slope.

$-8y = x + 32$ $x + 8y = 24$

Solve for y.

Write in slope-intercept form.

The slope of L_1 is _________. The slope of L_2 is _________.

Therefore, the lines are ____________________ because the slopes of the lines

are ____.

Name: Date:
Instructor: Section:

PRACTICE PROBLEMS

Write the equation of each line having the given slope and passing through the given point.

1. $m = \frac{1}{4}$, $(0,5)$ **1.**______________

2. $m = -4$, $(-3,7)$ **2.**______________

3. $m = \frac{-5}{8}$, $(-1,0)$ **3.**______________

Write the equation of each line passing through the given points.

4. $(5,4)$ and $(-7,2)$ **4.**______________

5. $(-6,1)$ and $(3,-4)$ **5.**______________

Determine if each pair of lines is parallel, perpendicular, or neither.

6. $L_1 : y = -\frac{1}{5}x + 4$ **6.**______________

$L_2 : y = -\frac{1}{5}x - 2$

7. $L_1 : y = \frac{4}{3}x - 7$ **7.**______________

$L_2 : y = -\frac{3}{4}x + 5$

Name: Date:
Instructor: Section:

8. $L_1: -2y = x - 4$ **8.**______________

$L_2: 2y = -x - 10$

9. $L_1: -2x + 3y = -9$ **9.**______________

$L_2: -x + 3y = 1$

Write the equation of each line passing through the given point with the given property. If the line is not horizontal or vertical, write the equation in standard form, $Ax + By = C$.

10. $(-3, 8)$ parallel to $y = -\frac{1}{4}x + 2$ **10.**______________

11. $(-1, -4)$ perpendicular to $-3x + 5y = 1$ **11.**______________

Write the equation of each line with the given conditions. If the line is not horizontal or vertical, write the equation in standard form, $Ax + By = C$.

12. Slope $m = \frac{1}{12}$ and y-intercept $(0, -4)$ **12.**______________

13. Vertical through the point $\left(-\frac{2}{3}, \frac{4}{5}\right)$ **13.**______________

14. Parallel to the x-axis containing the point $(-8, 7)$ **14.**______________

15. With x-intercept $(-2, 0)$ and y-intercept $(0, 3)$ **15.**______________

Name: Date:
Instructor: Section:

Chapter 3 LINEAR EQUATIONS, SLOPE, INEQUALITIES, AND INTRODUCTION TO FUNCTIONS

3.5 Graphing Linear Inequalities in Two Variables

KEY VOCABULARY

Term	Definition	Example
Linear inequality in two variables		

KEY PROPERTIES, PROCEDURES, OR STRATEGIES

Graphing a Linear Inequality in Two Variables

1.
2.
3.

NOTES

Name: Date:

Instructor: Section:

GUIDED EXAMPLES

On the coordinate plane, graph the solutions to the inequalities.

a) $y \le \frac{4}{5}x - 1$ **b)** $-2x + 3y > 6$

Solution

a) Graph the line $y = \frac{4}{5}x - 1$ as a _____________ line because any point on the line will be a solution to the inequality. The slope is _____ and the y-intercept is _____.

Use a test point to determine which portion of the plane to shade. Choose a point that is convenient but is *not* on the line. Test point: ___________________

$y \le \frac{4}{5}x - 1$

$(\quad) \le \frac{4}{5}(\quad) - 1$

Because the test point causes the original inequality to be __________, we shade the portion of the plane that _______________ contain the test point.

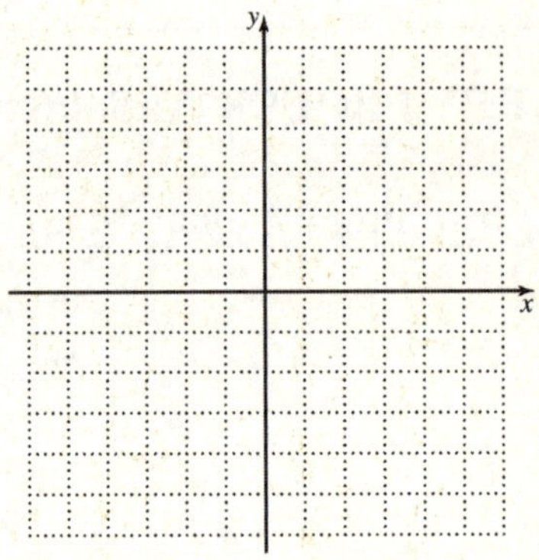

b) Graph the line $-2x + 3y = 6$ by any method. The line will be a ______________ line because the inequality is strict.

Solving the equation for y yields _________________. The line has slope ______ and y-intercept ______.

Select a test point not on the line: ______________.

$-2x + 3y > 6$

$-2(\quad) + 3(\quad) > 6$

Because the test point causes the original inequality to be __________, we shade the portion of the plane that _______________ contain the test point.

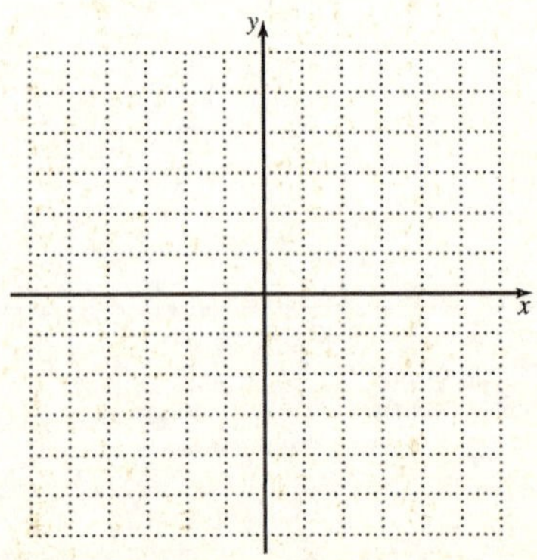

Name: Date:
Instructor: Section:

Chapter 3 LINEAR EQUATIONS, SLOPE, INEQUALITIES, AND INTRODUCTION TO FUNCTIONS

3.6 Introduction to Functions

KEY VOCABULARY

Term	Definition	Example
Relation		
Function		
Linear function		

NOTES

Name: Date:
Instructor: Section:

GUIDED EXAMPLES

1. Determine which of the following represent functions from the set $A=\{-4,\ -2,\ 0,\ 2,\ 4\}$ to the set $B=\{5,\ 6,\ 7,\ 8,\ 9,\ 10\}$.

a) $\{(-4,6),(-2,8),(0,6),(2,10),(4,5)\}$

b) $\{(0,8),(-2,6),(2,9),(-4,10),(0,7)\}$

Solution

a) $\{(-4,6),(-2,8),(0,6),(2,10),(4,5)\}$

This ____________ a function, even though the range value ________ is paired with two domain values, __________ and _________.

b) $\{(0,8),(-2,6),(2,9),(-4,10),(0,7)\}$

This ____________ a function. The domain value ________ is paired with two range values, __________ and _________.

2. Graph the linear function $h(x)=-\frac{3}{4}x-4$ and state the domain and range using interval notation.

Solution

Find three solution points by choosing three convenient values for x and substituting them into the linear function $h(x)=-\frac{3}{4}x-4$.

$h(x)=-\frac{3}{4}x-4$	$h(x)=-\frac{3}{4}x-4$	$h(x)=-\frac{3}{4}x-4$
$h(\ \ )=-\frac{3}{4}(\ \ )-4$	$h(\ \ )=-\frac{3}{4}(\ \ )-4$	$h(\ \ )=-\frac{3}{4}(\ \ )-4$

Plot the solutions found above and connect with a straightedge.

$D=$

$R=$

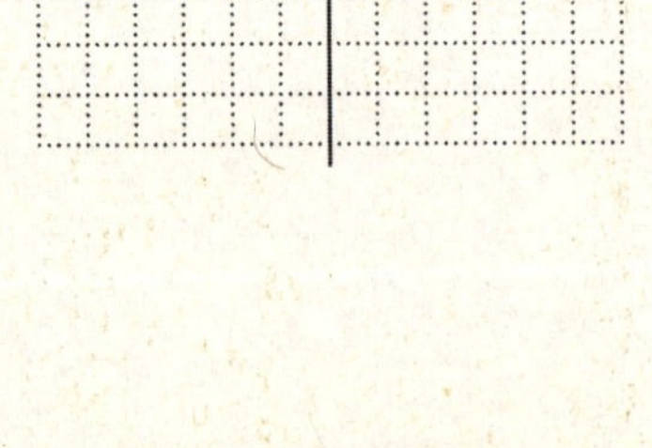

Name: Date:
Instructor: Section:

Chapter 4 LINEAR SYSTEMS OF EQUATIONS AND INEQUALITIES

4.1 The Graphing Method

KEY VOCABULARY

Term	Definition	Example
System of linear equations in two variables		
Consistent system		
Inconsistent system		
Independent system		
Dependent system		

KEY PROPERTIES, PROCEDURES, OR STRATEGIES

The Graphing Method

1. **2.** **3.** **4.**

Name: Date:
Instructor: Section:

GUIDED EXAMPLE

Solve the system of linear equations using the graphing method.

$$\begin{matrix} L_1 \\ L_2 \end{matrix} \begin{cases} -3x-2y=-1 \\ 2x+6y=-18 \end{cases}$$

Solution

Rewrite each equation in slope-intercept form $y=mx+b$ and determine its slope and y-intercept.

$L_1: -3x-2y=-1$

Slope:

y-intercept:

$L_2: 2x+6y=-18$

Slope:

y-intercept:

Now graph each line.

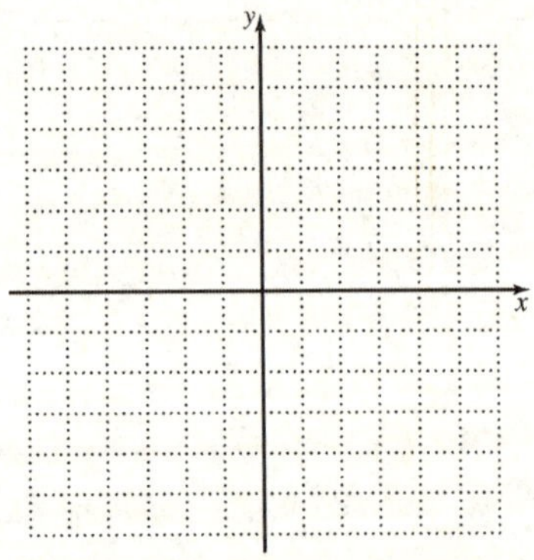

The solution set to the system is

Circle the correct choices:

The lines intersect, so the system is [consistent/inconsistent] and [dependent/independent].

Check the solution in each equation to see if it satisfies both of the original equations.

$-3(\quad)-2(\quad)=-1$

$2(\quad)+6(\quad)=-18$

Name: Date:
Instructor: Section:

Chapter 4 LINEAR SYSTEMS OF EQUATIONS AND INEQUALITIES

4.2 The Substitution Method

KEY PROPERTIES, PROCEDURES, OR STRATEGIES

The Substitution Method

1.

2.

3.

4.

5.

6.

NOTES

Name: Date:
Instructor: Section:

GUIDED EXAMPLE

Solve the system of linear equations using the Substitution Method. State whether the system is consistent/independent, consistent/dependent, or inconsistent.

$$\begin{cases} -4x+3y=1 \\ x+2y=8 \end{cases}$$

Solution

Solve the second equation for x, because its coefficient is 1.

$x+2y=8$

$x=$ ☐

Substitute this expression for x into the first equation and solve for y.

First equation: $-4x+3y=1$

☐ **Substitute for *x*.**

Solve for *y*.

Now find x by substituting this value for y into the equation we previously solved for x.

☐ **Substitute for *y*.**

Simplify.

The solution set is ☐

Check:

$-4x+3y=1$	$x+2y=8$
$-4(\quad)+3(\quad)=1$	$(\quad)+2(\quad)=8$

What type of system is this? ______________________________

Name: Date:
Instructor: Section:

Chapter 4 LINEAR SYSTEMS OF EQUATIONS AND INEQUALITIES

4.3 The Elimination Method

KEY PROPERTIES, PROCEDURES, OR STRATEGIES

The Elimination Method

1.

2.

3.

4.

5.

6.

7.

NOTES

Name: Date:
Instructor: Section:

GUIDED EXAMPLE

Solve the system of linear equations using the Elimination Method.

$$\begin{cases} -6x+3y=4 \\ 2x+4y=12 \end{cases}$$

Solution

The equations are written in standard form. Choose to eliminate the variable x. In order to get opposite coefficients on the variable x, multiply Equation 2 by 3.

Equation 1 $\begin{cases} -6x+3y=4 \\ 2x+4y=12 \end{cases}$
Equation 2

Equation 1 $-6x+3y=4$

3(Equation 2) [ ]

[ ]

$y =$

Scratch work for (3)(Eq 2)

$3(2x+4y=12)$

[ ]

Substitute into Equation 2 and solve for x.

$$2x+4y=12$$

$$2x+4(\quad)=12$$

[ ]

The solution set to the system of equations is [ ]

Check: Substitute the values for x and y into both original equations to verify the solution.

Name: Date:
Instructor: Section:

Chapter 4 LINEAR SYSTEMS OF EQUATIONS AND INEQUALITIES

4.4 Systems of Linear Equations in Three Variables

KEY VOCABULARY

Term	Definition	Example
Linear equation in three variables		

KEY PROPERTIES, PROCEDURES, OR STRATEGIES

Solving a Linear System in Three Variables

1.

2.

3.

4.

5.

6.

7.

Name: Date:
Instructor: Section:

GUIDED EXAMPLE

Solve the system of equations.

$$\begin{cases} x + y - z = 1 \\ 3x - 4y + z = 6 \\ -4x + 2y + 3z = 8 \end{cases}$$

Solution

Label the equations as follows:

$$\begin{matrix} \text{Eq 1} \\ \text{Eq 2} \\ \text{Eq 3} \end{matrix} \begin{cases} x + y - z = 1 \\ 3x - 4y + z = 6 \\ -4x + 2y + 3z = 8 \end{cases}$$

Let's keep Eq 1 because the coefficients are 1 or -1.

Choose to eliminate z from Eq 2 using Eq 1. Add Eq 1 and Eq 2 to obtain a new Eq 4.

Eq 1: $x + y - z = 1$

Eq 2: $3x - 4y + z = 6$

[ ] Eq 4

Choose to eliminate z from Eq 3 using Eq 1. Multiply Eq 1 by 3 and add to Eq 3 to obtain a new Eq 5.

(3)Eq 1: $3x + 3y - 3z = 3$

Eq 3: $-4x + 2y + 3z = 8$

[ ] Eq 5

Rewrite the system of equations: Eq 1 $x + y - z = 1$

Eq 4

Eq 5

Choose to keep either Eq 4 or Eq 5. In this example, we will keep Eq 5. Solve Eq 4 and Eq 5 to get an equation with only one unknown, a new Eq 6. This is the third equation for the new system.

[ ]

Now write the system so it can be solved using back substitution.

Eq 1 $x + y - z = 1$

Eq 5

Eq 6

Using back substitution, the solution set to the system is [ ]

Check the solutions in the original system.

Name: Date:
Instructor: Section:

PRACTICE PROBLEMS

Solve each system of linear equations.

1. $\begin{cases} -2x+3y+4z=12 \\ \quad\quad\;\; 4y+3z=2 \\ \quad\quad\quad\quad\;\; 4z=-8 \end{cases}$

1.________________

2. $\begin{cases} 2x+y-2z=2 \\ x-y+2z=4 \\ x+2y-z=0 \end{cases}$

2.________________

3. $\begin{cases} 3x+y-2z=2 \\ -3x-y+2z=-2 \\ 6x+2y-4z=4 \end{cases}$

3.________________

Name: Date:
Instructor: Section:

4. $\begin{cases} 3x+y-2z=2 \\ -x+y-2z=0 \\ 6x+2y-4z=5 \end{cases}$

4.________________

5. $\begin{cases} 2x-3y-2z=-8 \\ -4x+6y-8z=10 \\ 8x+12y-4z=10 \end{cases}$

5.________________

6. $\begin{cases} x+4y \qquad =-6 \\ \qquad 3y-5z=16 \\ x \qquad +2z=-4 \end{cases}$

6.________________

Name: Date:
Instructor: Section:

Chapter 4 LINEAR SYSTEMS OF EQUATIONS AND INEQUALITIES

4.5 Applications

GUIDED EXAMPLE

The sum of two integers is -4 and the difference between the larger and the smaller is 28. Find the two integers.

Solution

1 Prepare

We are asked to find two integers that satisfy the two conditions that their sum is -4 and their difference is 28.

2 Plan

Because there are two unknowns and two conditions, we can assign a different variable to each unknown integer and write a system of equations to represent the problem.

3 Process

Let x = the larger integer and y = the smaller integer. The sum of the two integers is -4. Use this information to write the first equation.

Eq 1: ______

The difference between the larger and the smaller integers is 28. Use this information to write the second equation.

Eq 2: ______

Set up the system of equations. Which method will be the most efficient to solve this system? Use this method to solve the system of equations.

The two integers are ______

4 Ponder

Do the integers satisfy the original conditions in the problem?

Name: Date:
Instructor: Section:

GUIDED EXAMPLE

Ashton wants to make 2 gallons of a 15% glucose solution by mixing an 8% glucose solution with a 20% glucose solution. How many gallons of each are needed?

Solution

1 Prepare

We are asked to determine the number of gallons of an 8% glucose solution and a 20% glucose solution that must be mixed to obtain 2 gallons of a 15% glucose solution.

2 Plan

There are two unknowns and two solutions, so we can set up a system of equations and solve by any method.

3 Process

Let x = number of gallons of an 8% glucose solution, and
y = number of gallons of a 20% glucose solution.

Use a chart to organize the information and help set up the equations.

	8% glucose	20% glucose	15% glucose
Amount (gallons)			
Percent Glucose			
Amount of Glucose			

Now use the chart to set up the equations.

Eq 1:

Eq 2:

Set up the system of equations. Which method will be the most efficient to solve this system? Use this method to solve the system of equations.

The amounts of each solution required are

4 Ponder

Do our answers seem reasonable?

Name: Date:
Instructor: Section:

Chapter 4 LINEAR SYSTEMS OF EQUATIONS AND INEQUALITIES

4.6 Systems of Linear Inequalities

KEY PROPERTIES, PROCEDURES, OR STRATEGIES

Solving a System of Linear Inequalities in Two Variables by Graphing

1.

2.

GUIDED EXAMPLES

1. Solve the system of linear inequalities by shading the appropriate region of the plane.

$$\begin{cases} x+6y \geq -12 \\ 5x-3y > -3 \end{cases}$$

Solution

Step 1 Rewrite each line in slope-intercept form $y = mx + b$ and determine its slope and y-intercept. Graph each line on the next page.

L_1: $x+6y=-12$

$m =$ ☐ and y-intercept ☐

Circle one: The line will be [solid/dashed].

L_2: $5x-3y=-3$

$m =$ ☐ and y-intercept ☐

Circle one: The line will be [solid/dashed].

To determine which half-plane to shade, select a test point for each linear inequality. You may use arrows to indicate the direction you would like to shade for each inequality's solution set.

L_1: $x+6y \geq -12$, test point ☐

Shade the appropriate region of the graph.

L_2: $5x-3y > -3$, test point ☐

Shade the appropriate region of the graph.

continued on the next page

Name: Date:
Instructor: Section:

Step 2 Shade the region of the plane that is the intersection of the two solution sets.

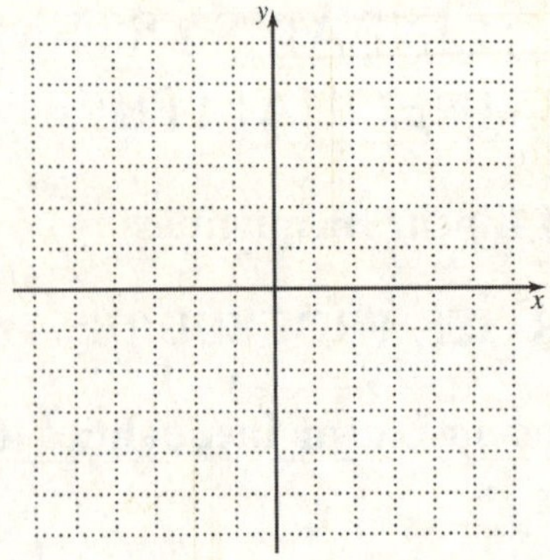

2. Solve the system of linear inequalities by shading the appropriate region of the plane.

$$\begin{cases} 3x+4y<8 \\ x-3y\leq 12 \end{cases}$$

Solution

Step 1 Rewrite each line in slope-intercept form $y=mx+b$ and determine its slope and y-intercept. Graph each line.

L_1: $3x+4y=8$

$m=$ ☐ and y-intercept ☐

Circle one: The line will be [solid/dashed].

L_2: $x-3y=12$

$m=$ ☐ and y-intercept ☐

Circle one: The line will be [solid/dashed].

To determine which half-plane to shade, select a test point for each linear inequality. You may use arrows to indicate the direction you would like to shade for each inequality's solution set.

L_1: $3x+4y<8$, test point ☐

Shade the appropriate region of the graph.

L_2: $x-3y\leq 12$, test point ☐

Shade the appropriate region of the graph.

Step 2 Shade the region of the plane that is the intersection of the two solution sets.

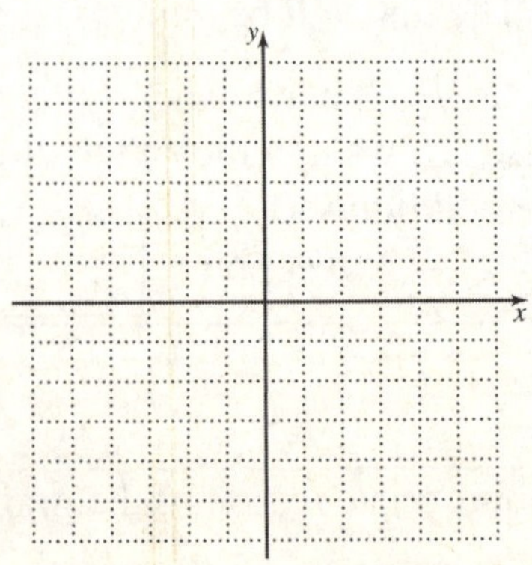

Name: Date:
Instructor: Section:

PRACTICE PROBLEMS

Solve each system of linear inequalities by shading the appropriate region in the plane.

1. $\begin{cases} x+4y \le 8 \\ 2x-3y \le 6 \end{cases}$

1.

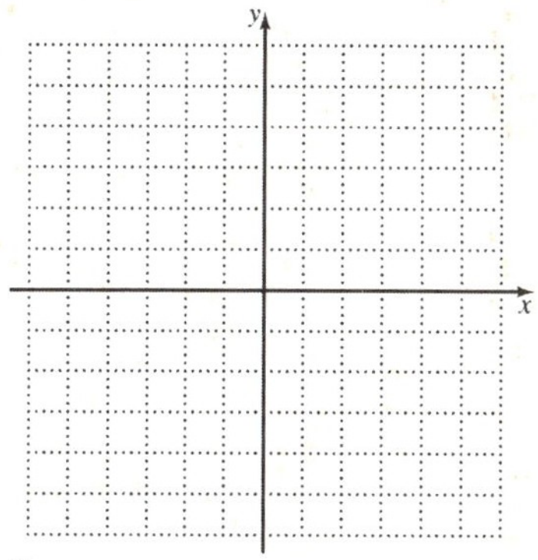

2. $\begin{cases} 2x+3y > -9 \\ 4x-5y < 25 \end{cases}$

2.

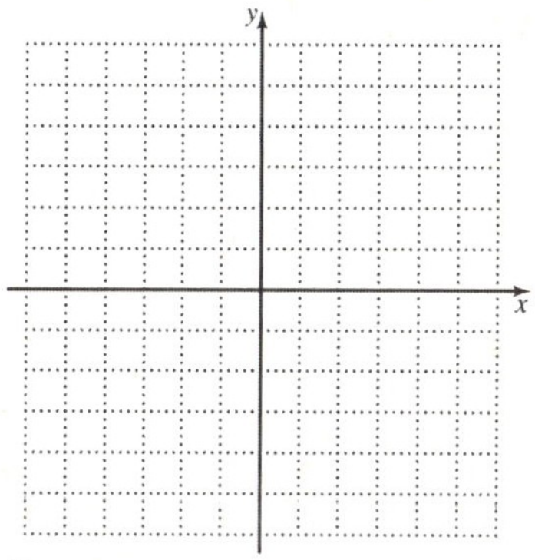

3. $\begin{cases} 5x+3y < 3 \\ -8x+5y \le 15 \end{cases}$

3.

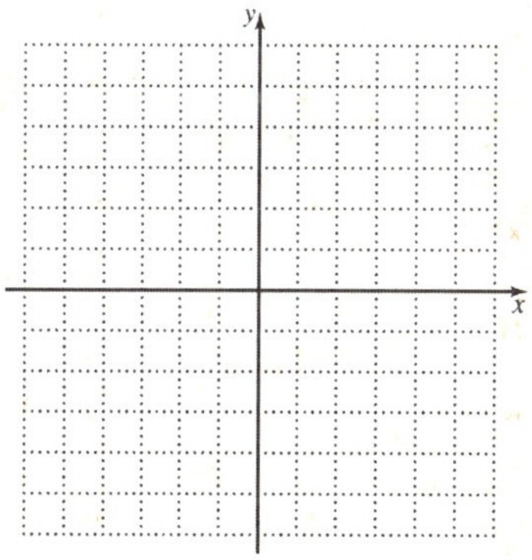

4. $\begin{cases} 4x+2y \ge 4 \\ 3x-2y > 6 \end{cases}$

4.

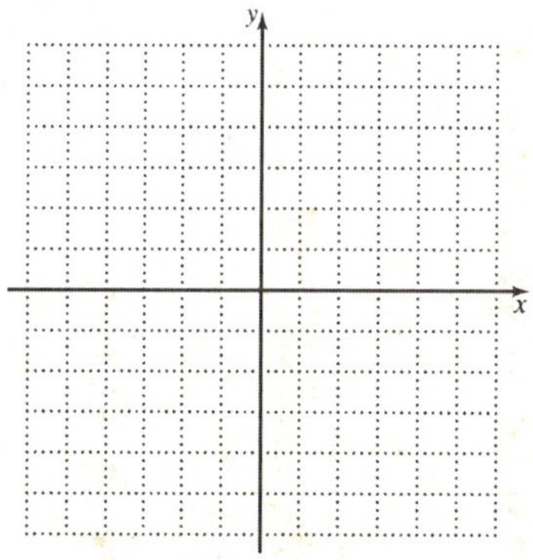

Name: Date:
Instructor: Section:

5. $\begin{cases} 2x-3y \geq -6 \\ 2x-3y < 9 \end{cases}$

5.

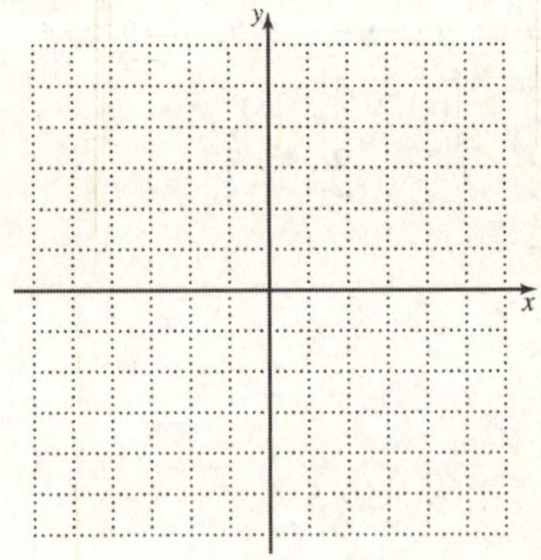

6. $\begin{cases} 5x+3y < 9 \\ 5x+3y \geq -6 \end{cases}$

6.

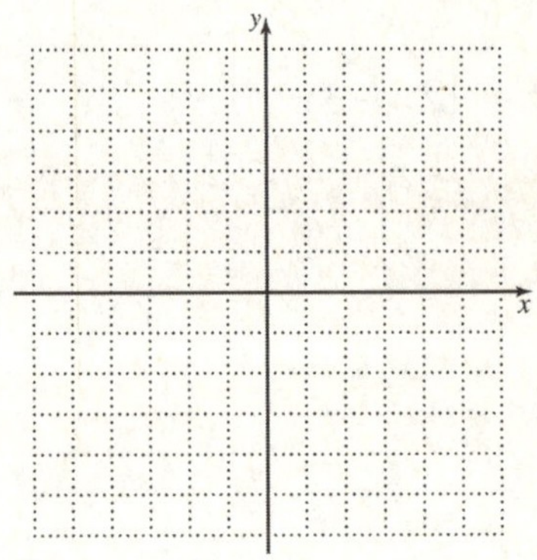

7. $\begin{cases} x \geq -4 \\ y > -1 \end{cases}$

7.

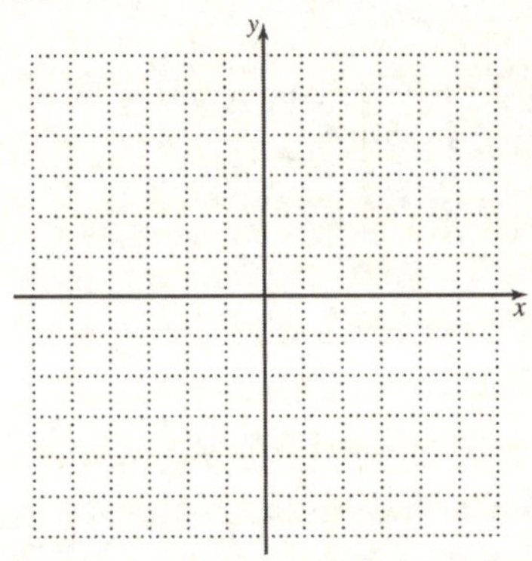

8. $\begin{cases} 3x-5y < -5 \\ y \leq 4 \\ x \geq -3 \end{cases}$

8.

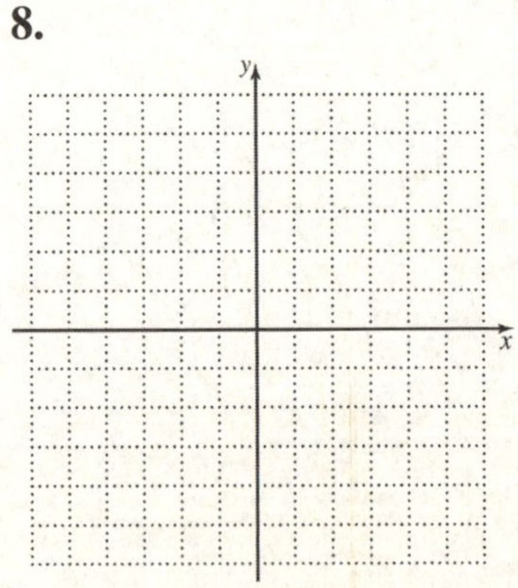

Name: Date:
Instructor: Section:

Chapter 5 EXPONENTS, POLYNOMIALS, AND APPLICATIONS

5.1 Integer Exponents

KEY PROPERTIES, PROCEDURES, OR STRATEGIES

Let x and y be real numbers such that $x \neq 0$, $y \neq 0$ and m and n are integers.

Product Rule

In the Language of Math	In Your Own Words

Power-to-Power Rule

In the Language of Math	In Your Own Words

Product-to-a-Power Rule

In the Language of Math	In Your Own Words

Quotient Rule

In the Language of Math	In Your Own Words

Quotient-to-a-Power Rule

In the Language of Math	In Your Own Words

Sign-Change Rule

In the Language of Math	In Your Own Words

Name: Date:
Instructor: Section:

GUIDED EXAMPLES

Assume that all variables represent nonzero real numbers. Simplify the following exponential expressions, writing answers in lowest terms with positive powers only.

a) $\dfrac{4^8 m^{-3}}{4^5 m^4}$ **b)** $\dfrac{\left(p^4 q^{-3}\right)^3}{p^{-2} q^5}$

Solution

a)

$\dfrac{4^8 m^{-3}}{4^5 m^4} = 4^{8-5} m^{-3-4}$ **Apply the** ______ **rule and subtract the exponents.**

$= 4^3 m^{-7}$ **Simplify the exponents.**

$=$ ______ **Apply the negative-exponent definition. Write as one fraction in lowest terms.**

b)

$\dfrac{\left(p^4 q^{-3}\right)^3}{p^{-2} q^5} = \dfrac{\left(p^4\right)^3 \left(q^{-3}\right)^3}{p^{-2} q^5}$ **Apply the** ______ **rule.**

$= \dfrac{p^{12} q^{-9}}{p^{-2} q^5}$ **Apply the** ______ **rule.**

$= \dfrac{p^{12} p^2}{q^5 q^9}$ **Apply the** ______ **rule.**

$= \dfrac{p^{14}}{q^{14}}$ **Apply the** ______ **rule.**

TES

Name: Date:
Instructor: Section:

Chapter 5 EXPONENTS, POLYNOMIALS, AND APPLICATIONS

5.2 Scientific Notation and Applications

KEY VOCABULARY

Term	Definition	Example
Scientific notation		

KEY PROPERTIES, PROCEDURES, OR STRATEGIES

To Convert from Standard Notation to Scientific Notation

1.

2.

To Convert from Scientific Notation to Standard Notation

1.

2.

3.

NOTES

Name: Date:
Instructor: Section:

GUIDED EXAMPLES

1. Write each number using scientific notation.

a) 807,565,000 **b)** 0.0000007506

Solution

a) 807,565,000

$= 8.07565\left(10^{\square}\right)$

Move _______ places left.

We move the decimal point to create a number greater than or equal to 1 but less than 10.

b) 0.000002674

$= 2.674\left(10^{\square}\right)$

Move _______ places right.

The number is very small and less than 1. Therefore, the exponent on the 10 is negative.

2. Simplify by first converting the number to scientific notation. Write the answer in both standard and scientific notation.

$(52,460,000)(0.0000000008)$

Solution

$(52,460,000)(0.0000000008) = 5.246\left(10^{\square}\right)\cdot 8.0\left(10^{\square}\right)$

7 places *10 places*

$= 41.968\left(10^{\square}\right)$ **Apply the product rule on the base 10.**

$= 4.1968\left(10^{\square}\right)\left(10^{\square}\right)$ **Write 41.968 in scientific notation**

$= 4.1968\left(10^{\square}\right)$ **Scientific notation**

$= \boxed{}$ **Standard notation**

Name: Date:
Instructor: Section:

Chapter 5 EXPONENTS, POLYNOMIALS, AND APPLICATIONS

5.3 Introduction to Polynomials and Polynomial Functions

KEY VOCABULARY

Term	Definition	Example
Degree of a term		
Polynomial		
Degree of a polynomial		

KEY PROPERTIES, PROCEDURES, OR STRATEGIES

Adding Polynomials

In the Language of Math	In Your Own Words

Subtracting Polynomials

In the Language of Math	In Your Own Words

Name: Date:
Instructor: Section:

GUIDED EXAMPLES

Perform the indicated operations.

a) $(4x^3 - 3x^2 + 2x - 7) + (5x^3 + 2x^2 - 6x + 2)$

b) $(2x^2 - 3x + 8) - (7x^2 + 5x - 3) + (2x - 1)$

Solution

a) $(4x^3 - 3x^2 + 2x - 7) + (5x^3 + 2x^2 - 6x + 2)$

$= 4x^3 - 3x^2 + 2x - 7 + 5x^3 + 2x^2 - 6x + 2$ — **Rewrite without parentheses.**

$= \underbrace{4x^3 + 5x^3} - \underbrace{3x^2 + 2x^2} + \underbrace{2x - 6x} - \underbrace{7 + 2}$ — **Use the Commutative and Associative Properties to rearrange the polynomial so that like terms are next to one another.**

[ ] [ ] [ ] [ ]

= [ ] — **Write the polynomial in standard form.**

b) $(2x^2 - 3x + 8) - (7x^2 + 5x - 3) + (2x - 1)$

= [ ] — **Remove parentheses and distribute the negative 1.**

= [ ] — **Group like terms.**

= [ ] — **Write in standard form.**

NOTES

Name: Date:
Instructor: Section:

Chapter 5 EXPONENTS, POLYNOMIALS, AND APPLICATIONS

5.4 Multiplying Polynomials

GUIDED EXAMPLES

1. Multiply $-4xy^2$ by $9x^4y^3$.

Solution

$\left(-4xy^2\right)\left(9x^4y^3\right)$

$=$ ☐ **Apply the Commutative and Associative Properties.**

$=$ ☐ **Apply the product rule and add exponents for like bases.**

$=$ ☐

2. Multiply the monomial $5a^3b^5$ by the polynomial $\left(-2ab+4a^2-b^4\right)$.

Solution

$5a^3b^5\left(-2ab+4a^2-b^4\right)$ **Translate into a mathematical expression; distribute the monomial to each term in the polynomial.**

$=$ ☐ **Multiply coefficients and apply the product rule for exponents.**

$=$ ☐

$=$ ☐ **Write in descending powers of the variable *a*.**

NOTES

Name: Date:
Instructor: Section:

GUIDED EXAMPLES

3. Multiply the binomials using the Distributive Property.
$(7x-4)(x+5)$

Solution
$(7x-4)(x+5)$

= [ ] **Distribute the binomial (*x* + 5).**

= [ ] **Distribute to remove parentheses, then combine like terms.**

= [ ] **Write in standard form.**

4. Multiply the binomial $(x+6)$ by the trinomial $(3x^2-4x+1)$.

Solution
$(7x-4)(3x^2-4x+1)$

= [ ] **Distribute the trinomial.**

= [ ] **Distribute to remove parentheses; combine like terms.**

= [ ] **Write in standard form.**

NOTES

Name: Date:
Instructor: Section:

PRACTICE PROBLEMS

Multiply the monomials.

1. $-4x^3(2x^2y)$ 1.________________

2. $(-3x^2y^2)(-4x^4y^6)$ 2.________________

3. $\left(-\frac{2}{5}ab^3c^4\right)(4a^3b^5c)$ 3.________________

Multiply.

4. $-4t^4(5-8t^2)$ 4.________________

5. $6y^3(4y^2+3y)$ 5.________________

6. $-2a^2b(-a^3+7ab^2)$ 6.________________

7. $\frac{3}{5}s^2\left(\frac{1}{2}r^2s-2r^4\right)$ 7.________________

Multiply using the Distributive Property.

8. $(x-2)(x-8)$ 8.________________

9. $(z+6)(z+9)$ 9.________________

Name: Date:
Instructor: Section:

10. $(7x-y)(4x-y)$ **10.**________________

11. $(3a+5b)(4a-2b)$ **11.**________________

Multiply.

12. $-4x^3\left(x^2-6x+8\right)$ **12.**________________

13. $-10ab^3\left(7a^2-2ab+3b^2\right)$ **13.**________________

14. $(a-5)\left(6a^2-ab+2b^2\right)$ **14.**________________

15. $\left(x^2-3xy+4y^2\right)\left(x^2+7xy-y^2\right)$ **15.**________________

Name: Date:
Instructor: Section:

Chapter 5 EXPONENTS, POLYNOMIALS, AND APPLICATIONS

5.5 Special Products

KEY PROPERTIES, PROCEDURES, OR STRATEGIES

FOIL Method

In the Language of Math	In Your Own Words

Squaring a Binomial

In the Language of Math	In Your Own Words

Product of the Sum and Difference of Two Terms

In the Language of Math	In Your Own Words

NOTES

Name: Date:

Instructor: Section:

GUIDED EXAMPLES

1. Use the FOIL method to multiply the binomials $(8x+4)(3x-5)$.

Solution

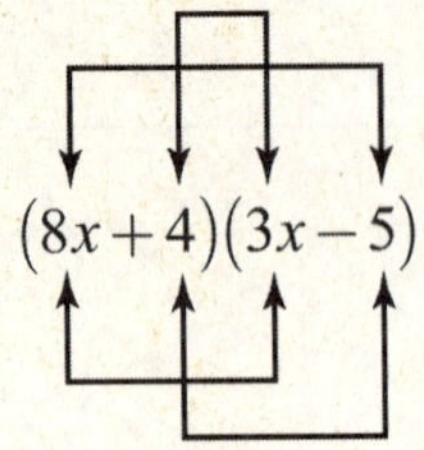

Identify the terms that correspond to the letters in FOIL.

$= \square + \square + \square + \square$

F O I L

Multiply using the FOIL method.

$= \square$

Simplify.

2. Square the binomials $(3y-2)^2$ using special products.

Solution

$(3y-2)^2 = \square + \square + \square$

First term squared **Twice the product of the two terms** **Last term squared**

$= \square$

3. Use the special product of a sum and difference of two terms to multiply the binomials $(6x-y)(6x+y)$.

Solution

$(6x-y)(6x+y) = \square - \square$

First term squared **Last term squared**

$= \square$

Name: Date:
Instructor: Section:

PRACTICE PROBLEMS

Multiply using the FOIL method.

1. $(4a-b)(3a+b)$ **1.**_______________

2. $(0.8x-y)(0.4x-6y)$ **2.**_______________

3. $(3a^2+2b^2)(-a^2-3b)$ **3.**_______________

Square using special products.

4. $(y+7)^2$ **4.**_______________

5. $(x-3z)^2$ **5.**_______________

6. $(4a+3b)^2$ **6.**_______________

7. $(8x^2-3y^2)^2$ **7.**_______________

Use the special product of a sum and difference of two terms to multiply the following binomials.

8. $(x-5)(x+5)$ **8.**_______________

Name: Date:
Instructor: Section:

9. $(7a-b)(7a+b)$ **9.**________________

10. $\left(\frac{2}{3}x+y\right)\left(\frac{2}{3}x-y\right)$ **10.**________________

11. $(x^2+6y^2)(x^2-6y^2)$ **11.**________________

Multiply.

12. $(4x+2y)^2$ **12.**________________

13. $(7a+3b)(2a-4b)$ **13.**________________

14. $\left(\frac{1}{5}x+\frac{1}{3}y\right)\left(\frac{1}{5}x-\frac{1}{3}y\right)$ **14.**________________

15. $(7x-3y)^2$ **15.**________________

Name: Date:
Instructor: Section:

Chapter 5 EXPONENTS, POLYNOMIALS, AND APPLICATIONS

5.6 Dividing Polynomials

REVIEW KEY VOCABULARY

Term	Definition	Example
Dividend		
Divisor		
Quotient		
Remainder		

KEY PROPERTIES, PROCEDURES, OR STRATEGIES

Long Division of Polynomials

1.
2.
3.
4.
5.
6.
7.

Name: Date:
Instructor: Section:

GUIDED EXAMPLES

1. Divide the polynomials.

$$\frac{-7x^3y^3 - 35x^2y^3}{-7xy^2}$$

Solution

$$\frac{-7x^3y^3 - 35x^2y^3}{-7xy^2} = \square + \square$$

Divide each term in the numerator by the denominator.

$$=$$

Simplify.

2. Use long division to divide $\frac{6x^3 - 13x^2 - 3x - 29}{x-3}$.

Solution

1st term of quotient:

$$x-3\overline{)6x^3 - 13x^2 - 3x - 29}$$

Divide. $\frac{6x^3}{x} = \square$

Multiply. $\square(x - 3) =$

Subtract. Bring down.

2nd term of quotient:

Divide. $\frac{\square}{x} = \square$

Multiply. $\square(x - 3) =$

Subtract. Bring down.

3rd term of quotient:

Divide. $\frac{\square}{x} = \square$

Multiply. $\square(x - 3) =$

Subtract. Bring down.

Answer: $\square$

Be sure to check your answer.

Name: Date:
Instructor: Section:

PRACTICE PROBLEMS

Divide. Do not use long division.

1. $\dfrac{8x^4 + 10x^3}{2x^3}$ **1.**________________

2. $\dfrac{-28x^5 + 56x^4 - 35x^3}{7x^2}$ **2.**________________

3. $\dfrac{3x^3y - x^2y^3 + 12xy^4}{3x^3y^4}$ **3.**________________

Divide using long division. Write the answers as Quotient + $\dfrac{\text{Remainder}}{\text{Divisor}}$.

4. $\dfrac{x^2 + 14x + 48}{x + 8}$ **4.**________________

5. $\dfrac{2x^3 - 16x^2 + 38x - 24}{x - 4}$ **5.**________________

Name: Date:
Instructor: Section:

6. $\dfrac{3x^3+5x^2-9x-18}{3x+5}$

6.________________

7. $\dfrac{4y^2+8y^3+12-2y}{2y+1}$

7.________________

8. $\dfrac{27x^3+27x^2+9x+1}{3x+1}$

8.________________

9. $\dfrac{8x^4-2x^3+x^2-6x}{x+1}$

9.________________

10. $\dfrac{-75x^3+260x^2-260x+80}{15x^2-40x+20}$

10.________________

Name: Date:
Instructor: Section:

Chapter 7 RATIONAL EXPRESSIONS

7.1 Simplifying Rational Expressions

KEY VOCABULARY

Term	Definition	Example
Rational expression		
Simplified rational expression		

KEY PROPERTIES, PROCEDURES, OR STRATEGIES

Simplifying a Rational Expression

1.

2.

NOTES

Name: Date:
Instructor: Section:

GUIDED EXAMPLES

Simplify each rational expression.

a) $\dfrac{4x+28}{2x^2-98}$

Solution

$\dfrac{4x+28}{2x^2-98}$

Factor completely and then reduce.

First factor out the GCF in each binomial.

$= \dfrac{\boxed{}}{\boxed{}}$

What is the GCF in the numerator? $\boxed{}$

What is the GCF in the denominator? $\boxed{}$

$= \dfrac{\boxed{}}{\boxed{}}$

Factor the binomial in the denominator.

Divide out any common factors.

$= \dfrac{\boxed{}}{\boxed{}}$

Write the denominator as a polynomial.

b) $\dfrac{9-49y^2}{7y-3}$

Solution

$\dfrac{9-49y^2}{7y-3}$

Factor completely and then reduce.

$= \dfrac{\boxed{}}{\boxed{}}$

Factor the binomial in the numerator. Note the *opposite* factors. Factor out a GCF = (-1) in the numerator.

Divide out any common factors.

$= \boxed{}$

Distribute the -1 to write the numerator as a polynomial.

Name: Date:
Instructor: Section:

Chapter 7 RATIONAL EXPRESSIONS

7.2 Multiplying and Dividing Rational Expressions

KEY PROPERTIES, PROCEDURES, OR STRATEGIES

Multiplying Rational Expressions

In the Language of Math	In Your Own Words

Dividing Rational Expressions

In the Language of Math	In Your Own Words

To Multiply and Divide Rational Expressions

1.

2.

3.

4.

NOTES

Name: Date:
Instructor: Section:

GUIDED EXAMPLES

Perform the indicated operations.

a) $\dfrac{a^2-9b^2}{3a^2+3ab+5ab+5b^2}\cdot\dfrac{3a^2+5ab}{a+3b}$

Solution

$$\frac{a^2-9b^2}{3a^2+3ab+5ab+5b^2}\cdot\frac{3a^2+5ab}{a+3b}$$

$$=\frac{\boxed{}}{\boxed{}}\cdot\frac{\boxed{}}{\boxed{}}$$

Factor all numerators and denominators completely. Divide out all common factors.

$$=\frac{\boxed{}}{\boxed{}}$$

Multiply the numerators and write it as a single polynomial.

b) $\dfrac{2ax-ay-2bx+by}{2a^2x-10ax+12x}\div\dfrac{b-a}{3x-5y}\cdot\dfrac{6x^3-24x}{6x^2-13xy+5y^2}$

Solution

$$\frac{2ax-ay-2bx+by}{2a^2x-10ax+12x}\div\frac{b-a}{3x-5y}\cdot\frac{6x^3-24x}{6x^2-13xy+5y^2}$$

$$=\frac{\boxed{}}{\boxed{}}\cdot\frac{\boxed{}}{\boxed{}}\cdot\frac{\boxed{}}{\boxed{}}$$

Write division as multiplication by the reciprocal.

$$=\frac{\boxed{}}{\boxed{}}\cdot\frac{\boxed{}}{\boxed{}}\cdot\frac{\boxed{}}{\boxed{}}$$

Factor all numerators and denominators completely. Divide out all common factors.

$$=\frac{\boxed{}}{\boxed{}}$$

Simplify.

Name: Date:

Instructor: Section:

PRACTICE PROBLEMS

Multiply each rational expression and write the answer in lowest terms.

1. $\dfrac{1a^5b^2}{3}\cdot\dfrac{9}{14a^4b^3}$ **1.**________________

2. $\dfrac{x^2-4y^2}{12x^2y}\cdot\dfrac{3xy^2}{x-2y}$ **2.**________________

3. $\dfrac{y^2-6y-16}{y^2-y-6}\cdot\dfrac{y^2-9}{y^2-5y-24}$ **3.**________________

Divide each rational expression and write the answer in lowest terms.

4. $\dfrac{15-5c}{25c^3d^7}\div\dfrac{c^3+3c^2}{10c^6d^4}$ **4.**________________

5. $\dfrac{12b^2+7b-12}{12b^2-29b+15}\div\dfrac{9b^2+27b+20}{9b^2-25}$ **5.**________________

6. $\dfrac{28-3x-x^2}{x^2+2x-8}\div\dfrac{x^2+3x-28}{2x^2+5x-12}$ **6.**________________

Name: Date:
Instructor: Section:

Perform the indicated operations and write the answers in lowest terms.

7. $\dfrac{x^2-16y^2}{18x^2y}\cdot\dfrac{2xy^2}{x-4y}$ **7.**______________

8. $\dfrac{24-2x-x^2}{x^2+2x-15}\div\dfrac{x^2+2x-24}{3x^2+13x-10}$ **8.**______________

9. $\dfrac{35c^4d^3}{3c^8}\cdot\dfrac{21c^4d}{14d^6}\div\dfrac{49c}{2d^3}$ **9.**______________

10. $\dfrac{6x^4y^2}{5x^7}\cdot\dfrac{10x^3y}{14y^4}\div\dfrac{4c}{7y^2}$ **10.**______________

Build up each rational expression to an equivalent rational expression with the indicated denominator.

11. $\dfrac{4a}{ac^4}=\dfrac{?}{4a^4b^4c^4}$ **11.**______________

12. $\dfrac{h-3}{h+6}=\dfrac{?}{(h+6)^2}$ **12.**______________

Name: Date:
Instructor: Section:

Chapter 7 RATIONAL EXPRESSIONS

7.3 Addition and Subtraction of Rational Expressions

KEY PROPERTIES, PROCEDURES, OR STRATEGIES

Adding and Subtracting Rational Expressions with Common Denominators

1.

2.

3.

4.

To Find the Least Common Denominator

1.

2.

To Add and Subtract Rational Expressions with Unlike Denominators

1.

2.

3.

4.

5.

6.

Name: Date:
Instructor: Section:

GUIDED EXAMPLE

Perform the indicated operation and write the answer in lowest terms.

$$\frac{3y}{y^2+2y-15}+\frac{8}{y^2-6y+9}$$

Solution

$$\frac{3y}{y^2+2y-15}+\frac{8}{y^2-6y+9}$$

$$=\frac{\square}{\square}+\frac{\square}{\square}$$

Factor each denominator completely.

$$=\frac{\square}{\square}+\frac{\square}{\square}$$

Build up each rational expression and write equivalent rational expressions with the LCD.

$$=\frac{\square}{\square}$$

Write as a single fraction.

$$=\frac{\square}{\square}$$

Combine like terms.

NOTES

Name: Date:
Instructor: Section:

PRACTICE PROBLEMS

Add or subtract each rational expression with common denominators. Write the answer in lowest terms.

1. $\dfrac{2a^2}{a^2-49}-\dfrac{11a+21}{a^2-49}$ **1.**________________

2. $\dfrac{2x^2+15x+5}{x^2+12x+32}+\dfrac{3x^2+7x+3}{x^2+12x+32}$ **2.**________________

3. $\dfrac{x^3}{x^2+8x+64}-\dfrac{512}{x^2+8x+64}$ **3.**________________

Find the LCD.

4. $\dfrac{3}{z}, \dfrac{2}{z-8}$ **4.**________________

5. $\dfrac{4}{x+6}, \dfrac{5}{x^2-36}$ **5.**________________

6. $\dfrac{x-4}{x+3}, \dfrac{x+9}{x-6}$ **6.**________________

Name: Date:
Instructor: Section:

Perform the indicated operations and write the answers in lowest terms.

7. $\dfrac{2c}{c+3}+\dfrac{7}{c}$ **7.**______________

8. $\dfrac{x+2}{x^2+9x+20}+\dfrac{6}{x^2+8x+15}$ **8.**______________

9. $\dfrac{y-4}{y-8}+\dfrac{4}{8-y}$ **9.**______________

10. $\dfrac{x^2}{x-12}+\dfrac{144}{12-x}$ **10.**______________

11. $\dfrac{8x}{x^2-9}-\dfrac{4}{x+3}-\dfrac{7}{3-x}$ **11.**______________

12. $\dfrac{x+8}{x^2-5x-14}-\dfrac{5x+3}{x^2+x-56}$ **12.**______________

Name: Date:
Instructor: Section:

Chapter 7 RATIONAL EXPRESSIONS

7.4 Synthetic Division and Complex Fractions

KEY VOCABULARY

Term	Definition	Example
Synthetic division		
Complex fraction		

KEY PROPERTIES, PROCEDURES, OR STRATEGIES

Method 1: Multiplying by the Reciprocal

1.

2.

3.

4.

Method 2: Multiplying by $\frac{\textbf{LCD}}{\textbf{LCD}}$

1.

2.

3.

4.

5.

Name: Date:
Instructor: Section:

GUIDED EXAMPLE

Use synthetic division to divide $\dfrac{2y^3+8y^2+7y-8}{y+2}$.

Solution

$\dfrac{2y^3+8y^2+7y-8}{y+2}$ is the same as $(2y^3+8y^2+7y-8)\div(y+2)$. Notice the divisor is not of the form $(y-k)$ and must be written as ☐ so that $k=$ ☐

Write the value of k and the coefficients of the dividend.

Bring down the leading coefficient.

Multiply and add to complete the synthetic division.

The quotient is ☐

The remainder is ☐

The final result is ☐

NOTES

Name: Date:
Instructor: Section:

Chapter 7 RATIONAL EXPRESSIONS

7.5 Solving Equations Containing Rational Expressions and Rational Functions

KEY VOCABULARY

Term	Definition	Example
Rational function		

KEY PROPERTIES, PROCEDURES, OR STRATEGIES

To Solve Rational Equations

1.

2.

3.

4.

5.

6.

NOTES

Name: Date:
Instructor: Section:

GUIDED EXAMPLES

1. Find any excluded values and state the domain of each the function using either interval or set-builder notation.

$$f(x)=\frac{x^2+2x-15}{x^2+4x-12}$$

Solution

$f(x)=\dfrac{x^2+2x-15}{x^2+4x-12}$. To find the excluded values, set the denominator equal to zero and solve for x.

[ ] = 0	**Set the denominator equal to zero.**
[ ] = 0	**Factor completely.**
[ ] = 0 or [ ] = 0	**Apply the Zero Factor Property.**
$x =$ [ ] $x =$ [ ]	

Therefore, the domain for f is $D =$ [ ]

2. Solve $\dfrac{6x}{x+2}+\dfrac{2}{x-3}=6.$

Solution

$\dfrac{6x}{x+2}+\dfrac{2}{x-3}=6.$

Find the LCD: [ ]

Multiply both sides of the equation by the LCD.

Distribute the LCD on the LHS. Write the RHS as a polynomial.

Set the equation equal to zero.

Solve for x.

Solution Set: [ ]

Name: Date:
Instructor: Section:

PRACTICE PROBLEMS

Find any excluded values and state the domain of each rational function using either interval or set-builder notation.

1. $f(t) = \dfrac{3t}{6+t} + 8$ **1.**________________

2. $f(x) = \dfrac{2x}{x^2+16}$ **2.**________________

3. $h(x) = \dfrac{4x^3}{x^2+2x-15}$ **3.**________________

Solve each equation.

4. $\dfrac{2}{x} + \dfrac{1}{7} = \dfrac{8}{x}$ **4.**________________

5. $\dfrac{3}{y+5} - \dfrac{1}{5-y} = \dfrac{2}{y^2-25}$ **5.**________________

6. $\dfrac{3x}{x+2} + \dfrac{4}{x-1} = 3$ **6.**________________

Name: Date:
Instructor: Section:

7. $\frac{9}{x+2}-\frac{2}{x-1}=\frac{-20}{x^2+x-2}$ **7.**________________

Identify whether each of the following is an algebraic expression or equation. If it is an algebraic expression, perform the indicated operations and write the answer in lowest terms. If it is an equation, solve it for the indicated variable.

8. $\frac{x+7}{7x}+\frac{4-x}{x^2}$ **8.**________________

9. $\frac{x-5}{x+7}=\frac{7}{2x-5}$ **9.**________________

Solve each formula for the indicated variable.

10. $\frac{1}{s}+\frac{1}{t}=\frac{1}{a}$ for a **10.**________________

11. $\frac{P_1V_1}{T_1}=\frac{P_2V_2}{T_2}$ for T_1 **11.**________________

Solve the following application algebraically.

12. The sum of the reciprocals of two consecutive integers is $\frac{7}{12}$. Find the two integers. **12.**________________

Name: Date:
Instructor: Section:

Chapter 7 RATIONAL EXPRESSIONS

7.6 Applications and Variations

KEY VOCABULARY

Term	Definition	Example
y varies directly as x, or y is directly proportional to x		
Variation constant		
y varies inversely with x, or y is inversely proportional to x		
y varies jointly with x and z, or y is jointly proportional to x and z		
y varies jointly with x and z, and inversely with w		

KEY PROPERTIES, PROCEDURES, OR STRATEGIES

To Solve Variation Problems

1.

2.

3.

4.

Name: Date:
Instructor: Section:

GUIDED EXAMPLE
Amber can rake and bag the leaves in her yard in 6 hours. It takes her brother Colin 5 hours to do the same job. How long would it take them to complete the job if they work together?

Solution

1. **Prepare**
   We will use the work equation

$$\begin{pmatrix}\text{Time}\\\text{Worked}\end{pmatrix}\begin{pmatrix}\text{Individual's}\\\text{Rate}\end{pmatrix}+\begin{pmatrix}\text{Time}\\\text{Worked}\end{pmatrix}\begin{pmatrix}\text{Individual's}\\\text{Rate}\end{pmatrix}+\cdots=1\text{ complete job}$$

   and Amber's and Colin's individual work rates to determine how long it would take them to rake and bag the leaves if they work together.

2. **Plan**
   Amber's time alone = 6 hours, so Amber's rate = $\square$ of the job per hour.

   Colin's time alone = 5 hours, so Colin's rate = $\square$ of the job per hour.

   Let x = time Amber and Colin work together.

3. **Process**
   Write the equation.

$$\begin{pmatrix}\text{Time}\\\text{Worked}\end{pmatrix}\begin{pmatrix}\text{Amber's}\\\text{Rate}\end{pmatrix}+\begin{pmatrix}\text{Time}\\\text{Worked}\end{pmatrix}\begin{pmatrix}\text{Colin's}\\\text{Rate}\end{pmatrix}=1$$

$$x\left(\square\right)+x\left(\square\right)=1$$

$$\square\, x\left(\square\right)+\square\, x\left(\square\right)=1\left(\square\right)$$

Clear fractions by multiplying both sides by the LCD = $\square$

Solve for *x*.

It would take Amber and Colin $\square$ hours, working together.

4. **Ponder**
   Does the answer seem reasonable?

Name: Date:
Instructor: Section:

PRACTICE PROBLEMS

1. Translate the statement.
G varies jointly with a and b and inversely with c.

1.________________

2. Translate the statement and find the variation constant, k.
C varies inversely with d, and $c = 2$ when $d = 6$.

2.________________

Solve each variation problem.

3. If w varies directly with b, and $w = 7$ when $b = 5$, find w when $b = 15$.

3.________________

4. A varies jointly with b and c; $A = 462$ when $b = 11$ and $c = 21$. If $b = 15$ and $c = 21$, find A.

4.________________

Solve.

5. Your paycheck (before deductions) varies directly as the number of hours you work. If your paycheck is \$193.80 for 19 hours, find your pay for 10 hours.

5.________________

Name: Date:
Instructor: Section:

6. The height of a triangle varies directly with its area and inversely with its base. The height of a triangle is 10 mm when its area is 30 mm^2 and the base is 6 mm. Find the height of a triangle with an area of 63 mm^2 and a base of 7 mm.

6.________________

Solve each problem by utilizing the 4P's: Prepare, Plan, Process, and Ponder.

7. John drove 189 miles to visit his girlfriend, Jane, for the Thanksgiving holidays. When he returned on Sunday, he followed Jane's suggested shortcuts and traveled only 168 miles, but his average speed was 5 mph slower. If the time traveled was the same each way, what was his average speed in each direction?

7.________________

8. Tracy can mow and edge the Kirks', Peters', and Wilsons' lawns in 3 hours alone, while Julia can do the same job in 8 hours. How long will it take them to mow and edge the three lawns if they work together?

8.________________

Name: Date:
Instructor: Section:

Chapter 8 ROOTS AND RADICALS

8.1 Simplifying, Adding, and Subtracting Square Roots

KEY VOCABULARY

Term	Definition	Example
Principal square root of *a*		
Square root of a perfect square		

KEY PROPERTIES, PROCEDURES, OR STRATEGIES

Simplified Radical Form

In the Language of Math	In Your Own Words
1. **2.** **3.**	

Product Rule for Radicals

In the Language of Math	In Your Own Words

Name: Date:
Instructor: Section:

GUIDED EXAMPLE

Perform the indicated operations. Write the answer in simplified radical form.

$$4\sqrt{2}+3\sqrt{50}-7\sqrt{8}$$

Solution

$4\sqrt{2}+3\sqrt{50}-7\sqrt{8}$ — **Note that 50 and 8 have perfect squares as factors. Write all terms as simplified radicals.**

$=4\sqrt{2}+3\sqrt{(\quad)(\quad)}-7\sqrt{(\quad)(\quad)}$ — **Factor all radicands, writing perfect squares first.**

$=4\sqrt{2}+3\sqrt{(\quad)}\sqrt{(\quad)}-7\sqrt{(\quad)}\sqrt{(\quad)}$ — **Apply the product rule for radicals.**

$=4\sqrt{2}+3(\quad)\sqrt{(\quad)}-7(\quad)\sqrt{(\quad)}$ — **Simplify the square roots.**

$=4\sqrt{2}+(\quad)\sqrt{(\quad)}-(\quad)\sqrt{(\quad)}$ — **Multiply and write as simplified radicals.**

$=(\qquad)\sqrt{(\quad)}$ — **Combine like radicals using the Distributive Property.**

$=(\quad)\sqrt{(\quad)}$ — **Write in simplified radical form.**

NOTES

Name: Date:
Instructor: Section:

PRACTICE PROBLEMS

Find the principal square roots.

1. $\sqrt{100}$ **1.**________________

2. $-\sqrt{\frac{64}{25}}$ **2.**________________

Simplify using the product rule for radicals.

3. $\sqrt{48}$ **3.**________________

4. $\sqrt{245}$ **4.**________________

Simplify. Assume that all variables represent real numbers.

5. $\sqrt{(8x)^2}$ **5.**________________

6. $-\sqrt{64b^2}$ **6.**________________

7. $\sqrt{a^2-12a+36}$ **7.**________________

Write in simplified radical form. Assume that all variables represent positive real numbers.

8. $\sqrt{y^{14}}$ **8.**________________

9. $\sqrt{z^{19}}$ **9.**________________

Name: Date:
Instructor: Section:

10. $\sqrt{48c^{51}}$ 10.________________

11. $\sqrt{192m^{15}n^{17}}$ 11.________________

Perform the indicated operations. Write the answers in simplified radical form.

12. $7\sqrt{2}-5\sqrt{2}+6\sqrt{2}$ 12.________________

13. $8\sqrt{5}-10\sqrt{5}+4\sqrt{7}$ 13.________________

14. $\sqrt{2}+7\sqrt{8}$ 14.________________

15. $4\sqrt{12}+5\sqrt{48}-2\sqrt{75}$ 15.________________

Name: Date:
Instructor: Section:

Chapter 8 ROOTS AND RADICALS

8.2 Multiplying and Dividing Square Roots

KEY PROPERTIES, PROCEDURES, OR STRATEGIES

Quotient Rule for Radicals

In the Language of Math	In Your Own Words

Rationalizing the Denominator

In the Language of Math	In Your Own Words

Multiply by the Conjugate

In the Language of Math	In Your Own Words

NOTES

Name: Date:

Instructor: Section:

GUIDED EXAMPLE

Rationalize the denominator. Write the answer in simplified radical form. Assume that all variables represent positive real numbers.

$$\sqrt{\frac{14x^4}{27y^7}}$$

Solution

$$\sqrt{\frac{14x^4}{27y^7}}$$

The radical is not in simplified form because there is a fraction underneath the radical and perfect squares as factors of the radicand.

$$=\frac{\sqrt{(\qquad)}}{\sqrt{(\qquad)}}$$

Apply the quotient rule for radicals to eliminate the fraction under the radical.

$$=\frac{\sqrt{(\qquad)(\qquad)}}{\sqrt{(\qquad)(\qquad)}}$$

Factor each radicand, writing the perfect squares first.

$$=\frac{(\qquad)\sqrt{(\qquad)}}{(\qquad)\sqrt{(\qquad)}}$$

Write the numerator and denominator in simplified form using the product rule for radicals.

$$=\frac{(\qquad)\sqrt{(\qquad)}}{(\qquad)\sqrt{(\qquad)}}\cdot\frac{(\qquad)}{(\qquad)}$$

Multiply by the special form of 1 needed to rationalize the denominator.

$$=\frac{(\qquad)\sqrt{(\qquad)}}{(\qquad)}$$

Write in simplified radical form.

Name: Date:
Instructor: Section:

PRACTICE PROBLEMS

Multiply. Write the answers in simplified radical form. Assume that all variables represent positive real numbers.

1. $\sqrt{5}\cdot\sqrt{75}$ **1.**________________

2. $\sqrt{6}\cdot\sqrt{15}$ **2.**________________

3. $\sqrt{10x^3}\cdot\sqrt{35x^2}$ **3.**________________

4. $\left(\sqrt{7y}\right)^2$ **4.**________________

Multiply. Write the answers in simplified radical form. Assume that all variables represent positive real numbers.

5. $\sqrt{3}\left(\sqrt{3}+\sqrt{5}\right)$ **5.**________________

6. $\sqrt{a}\left(\sqrt{a}+\sqrt{7}\right)$ **6.**________________

7. $\left(\sqrt{2}-\sqrt{10}\right)\left(\sqrt{15}+\sqrt{3}\right)$ **7.**________________

8. $\left(4+\sqrt{12}\right)^2$ **8.**________________

Name: Date:
Instructor: Section:

Divide. Write the answers in simplified radical form. Assume that all variables represent positive real numbers.

9. $\dfrac{\sqrt{3}}{\sqrt{48}}$ **9.**_______________

10. $\dfrac{\sqrt{21x^3}}{\sqrt{7x}}$ **10.**_______________

11. $\dfrac{\sqrt{98a^{10}}}{\sqrt{2a^4}}$ **11.**_______________

Rationalize each denominator. Write the answers in simplified radical form. Assume that all variables represent positive real numbers.

12. $\dfrac{8}{\sqrt{3}}$ **12.**_______________

13. $\dfrac{\sqrt{5x^2}}{\sqrt{18y^3}}$ **13.**_______________

14. $\dfrac{\sqrt{16}}{\sqrt{2}+\sqrt{21}}$ **14.**_______________

Name: Date:
Instructor: Section:

Chapter 8 ROOTS AND RADICALS

8.3 Solving Equations Containing Square Roots

KEY VOCABULARY

Term	Definition	Example
Radical equation		

KEY PROPERTIES, PROCEDURES, OR STRATEGIES

Squaring Property of Equality

In the Language of Math	In Your Own Words

To Solve Equations Containing Square Roots

1. **2.** **3.** **4.** **5.**	

NOTES

Name: Date:
Instructor: Section:

GUIDED EXAMPLES

Solve and write the solutions using set notation.

a) $\sqrt{x+12}=11$

Solution

$\sqrt{x+12}=11$ **The radical is isolated on the LHS.**

Apply the Squaring Property of Equality.

Solve the linear equation.

Check:

Solution Set: $\{\quad\}$

b) $\sqrt{x+8}-\sqrt{3x-4}=0$

Solution

$\sqrt{x+8}-\sqrt{3x-4}=0$

$$\sqrt{x+8}=\sqrt{3x-4}$$

$$\left(\sqrt{x+8}\right)^2=\left(\sqrt{3x-4}\right)^2$$

$$x+8=3x-4$$

$$12=2x$$

$$6=x$$

Check: $\sqrt{x+8}-\sqrt{3x-4}=0,\ x=6$

Solution Set: $\{\quad\}$

Name: Date:
Instructor: Section:

PRACTICE PROBLEMS

Solve and write the solutions using set notation.

1. $\sqrt{x+17}=12$ **1.**________________

2. $\sqrt{4x-20}=14$ **2.**________________

3. $9+\sqrt{12x}=27$ **3.**________________

4. $\sqrt{3x+8}+7=0$ **4.**________________

5. $\sqrt{16x-8}=\sqrt{9x+3}$ **5.**________________

Name: Date:
Instructor: Section:

6. $\sqrt{10x^2+4x}=\sqrt{5x}$

6.________________

7. $\sqrt{11-x}=x-9$

7.________________

8. $\sqrt{7+x}=\sqrt{8x^2}$

8.________________

9. $\sqrt{x}+4=\sqrt{7+18x}$

9.________________

10. $\sqrt{5x^2}=\sqrt{2x-x^2}$

10.________________

Name: Date:
Instructor: Section:

Chapter 8 ROOTS AND RADICALS

8.4 Applications

KEY PROPERTIES, PROCEDURES, OR STRATEGIES

Properties of a 45°-45°-90° Triangle

Properties of a 30°-60°-90° Triangle

Distance Between Two Points

Name: Date:
Instructor: Section:

GUIDED EXAMPLE

The length of the hypotenuse of a 45°-45°-90° triangle is $8\sqrt{3}$ inches. Find the length of the legs.

Solution

The length of the hypotenuse in this type of triangle will always be $\sqrt{2}$ times the length of a leg. So, given the length of the hypotenuse, the length of a leg can be found by

The length of the hypotenuse is $8\sqrt{3}$. Then the length of each leg is

Simplify this expression.

Therefore, the length of each leg is

NOTES

Name: Date:
Instructor: Section:

PRACTICE PROBLEMS

Using the distance formula, find the distance between the points.

1. $(-3,5)$ and $(4,-2)$ **1.**________________

2. $(8,-5)$ and $(-1,6)$ **2.**________________

3. $(0,-8)$ and $(7,9)$ **3.**________________

Solve by using an algebraic equation.

4. The square root of the sum of a number and 6 is equal to the square root of five times the number. Find the number(s). **4.**________________

5. The square root of six times a number is equal to the square root of the sum of 8 and the number. Find the number(s). **5.**________________

Name: Date:
Instructor: Section:

6. The square root of seven less than three times a number is equal to the square root of the sum of 4 plus six times the number. Find the number(s).

6.________________

7. The square root of the sum of a number and 5 is equal to the sum of the number and 5. Find the number(s).

7.________________

8. The sum of a number and 8 is the same as the square root of the sum of three times the number and 24. Find the number(s).

8.________________

9. The length of a leg of a 45°-45°-90° triangle is $x = 4\sqrt{14}$ meters. Find the length of the hypotenuse.

9.________________

10. The hypotenuse of a 30°-60°-90° triangle measures $8\sqrt{5}$ feet. Find the lengths of the legs.

10.________________

Name: Date:
Instructor: Section:

Chapter 8 ROOTS AND RADICALS

8.5 Higher Roots and Rational Exponents

KEY VOCABULARY

Term	Definition	Example
Principal *n*th root		

KEY PROPERTIES, PROCEDURES, OR STRATEGIES

Product Rule and Quotient Rule for Radicals

In the Language of Math	In Your Own Words

Rational Exponents

In the Language of Math	In Your Own Words

Radical Expression with a Rational Exponent $\frac{m}{n}$

In the Language of Math	In Your Own Words

Name: Date:
Instructor: Section:

GUIDED EXAMPLES

1. Simplify the expression using the product rule. Write the answer in simplified radical form. Assume that all variables represent positive real numbers.

$\sqrt[3]{32a^5b^6}$

Solution

$\sqrt[3]{32a^5b^6}$ **Because the index is 3, think about perfect cubes.**

$=\sqrt[3]{(\quad)\cdot(\quad)}$ **Write perfect powers of 3 first.**

$=\sqrt[3]{(\quad)}\cdot\sqrt[3]{(\quad)}$ **Apply the product rule for radicals.**

$=(\quad)\sqrt[3]{\quad}$ **Simplify the cube root and write in simplified radical form.**

2. Using the exponent rules, simplify the expression. Write the expression with positive exponents only. Assume that all variables represent positive real numbers.

$\left(27y^{-\frac{3}{4}}\right)^{-\frac{5}{3}}$

Solution

$\left(27y^{-\frac{3}{4}}\right)^{-\frac{5}{3}} = \boxed{}$ **Apply the product-to-a-power rule.**

$= \boxed{}$ **Apply the definition of negative exponents.**

$= \boxed{}$ **Simplify the *n*th root.**

Name: Date:
Instructor: Section:

Chapter 8 ROOTS AND RADICALS

8.6 Radical Functions, Equations Containing Higher Roots

KEY VOCABULARY

Term	Definition	Example
Radical function		
Radical equation		

KEY PROPERTIES, PROCEDURES, OR STRATEGIES

Power Rule

In the Language of Math	In Your Own Words

To Solve Equations Containing Radicals (Higher Roots)

1.

2.

3.

4.

5.

Name: Date:
Instructor: Section:

GUIDED EXAMPLES

Solve and write the solutions using set notation.

a) $\sqrt[3]{8x-5}-6=-8$

Solution

$\sqrt[3]{8x-5}-6=-8$

Isolate the radical on the LHS by adding 6 to both sides.

Apply the power rule and raise each side to a power of 3.

Solve the linear equation for *x*.

Check:

Solution Set:

b) $\sqrt[4]{5x-6}-\sqrt[4]{3x-2}=0$

Solution

$\sqrt[4]{5x-6}-\sqrt[4]{3x-2}=0$

Isolate the radicals on each side of the equation.

Apply the power rule and raise each side to a power of 4.

Gather variables to the LHS and solve.

Check:

Solution Set:

Name: Date:
Instructor: Section:

Chapter 9 FUNCTIONS

9.1 More on Functions

REVIEW KEY VOCABULARY

Term	Definition	Example
Function		

KEY PROPERTIES, PROCEDURES, OR STRATEGIES

The Vertical Line Test for Functions

NOTES

Name: Date:
Instructor: Section:

GUIDED EXAMPLES

1. Find the value of $f(x-4)$ for $f(x)=3x^2+5x-7$.

Solution

$f(x)=3x^2+5x-7$, $f(x-4)$

$f(\quad)=3(\quad)^2+5(\quad)-7$ — **Substitute in $(x-4)$ for every x in the function f.**

Simplify.

2. Applying the vertical line test, determine if the following graph represents a function.

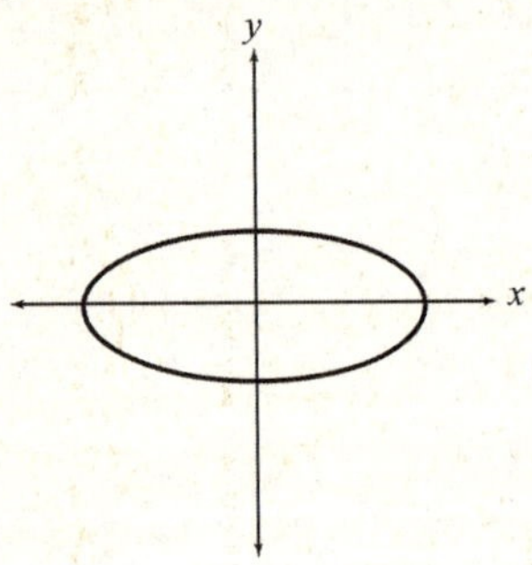

Solution

According to the vertical line test for functions, the graph represents a function if and only if

Can you draw a vertical line that intersects the graph more than once? If so, draw such a line.

Circle one: This graph [does/does not] represent the graph of a function.

Name: Date:
Instructor: Section:

Chapter 9 FUNCTIONS

9.2 The Algebra of Functions

KEY PROPERTIES, PROCEDURES, OR STRATEGIES

The Algebra of Functions

In the Language of Math	In Your Own Words

Composition of Functions

In the Language of Math	In Your Own Words

Name: Date:
Instructor: Section:

GUIDED EXAMPLE

Let $g(x)=\dfrac{5}{x+4}$ and $h(x)=\dfrac{3}{x^2-16}$. Find $\left(\dfrac{g}{h}\right)(x)$ and its domain.

Solution

$\left(\dfrac{g}{h}\right)(x)=$ [ ] **State the definition of $\left(\dfrac{g}{h}\right)(x)$.**

[ ] **Substitute the functions for g and h.**

[ ] **Rewrite without the fraction bar.**

[ ] **Multiply by the reciprocal and factor.**

[ ] **Simplify.**

The domain of g is [ ]

The domain of h is [ ]

The domain of the quotient is [ ]

NOTES

Name: Date:
Instructor: Section:

Chapter 9 FUNCTIONS

9.4 Translations

KEY PROPERTIES, PROCEDURES, OR STRATEGIES

Graphs of the Six Basic Functions

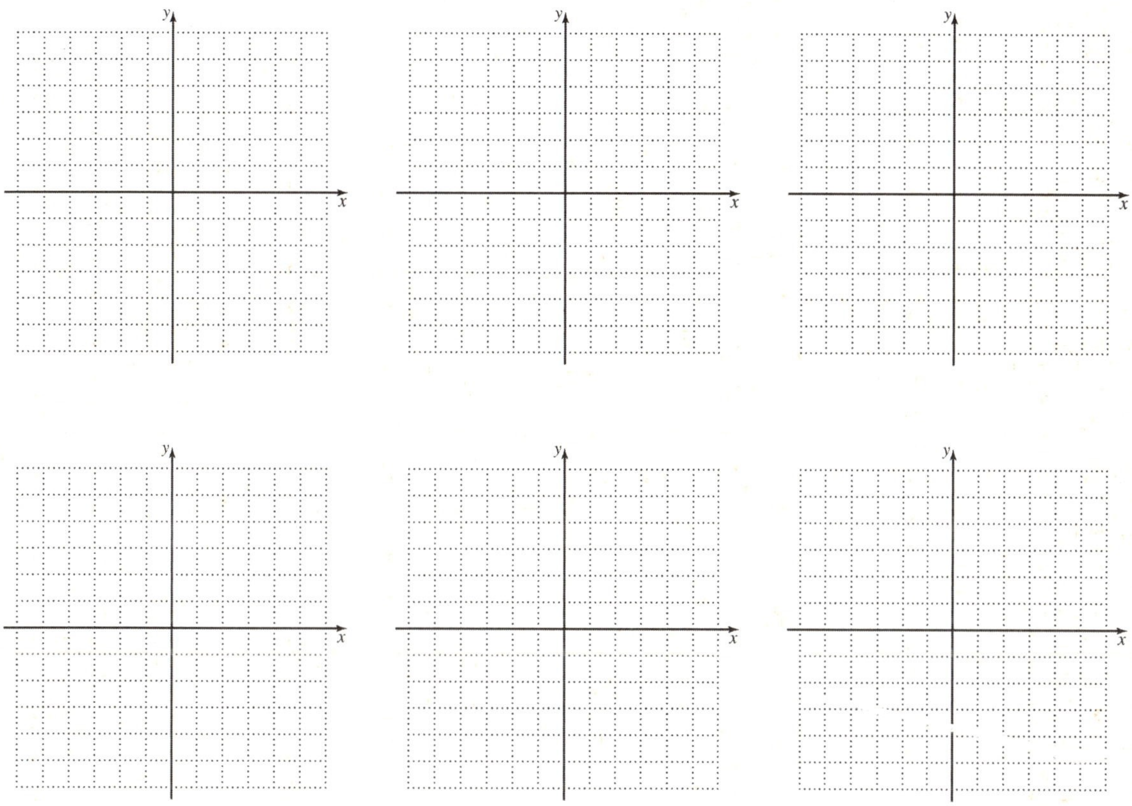

Translations (Rigid Transformations)

1.

2.

3.

4.

Name: Date:
Instructor: Section:

Reflections about the Axes

1.

2.

GUIDED EXAMPLE

Using translations and one of the basic graphs, sketch the function $f(x) = -|x+2| - 3$. State the translations.

Solution

The original function is

The new function $f(x) = -|x+2| - 3$ is a combination of three translations. Remember, the *order* in which we apply the transformations is important. Keep in mind the order of operations.

1st:

2nd:

3rd:

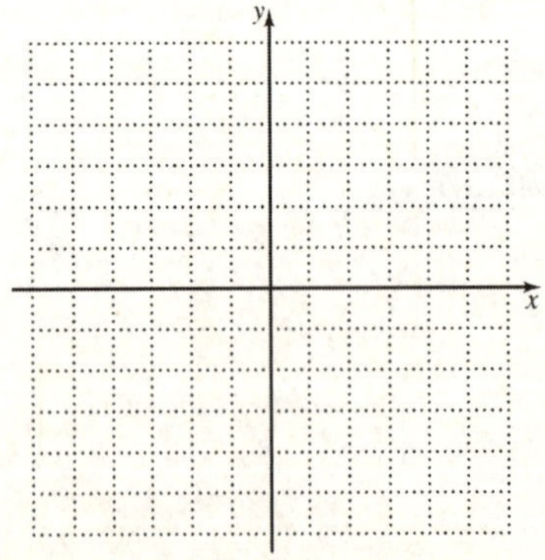

Name: Date:
Instructor: Section:

PRACTICE PROBLEMS

Using transformations and/or reflections and one of the basic graphs, state the transformation and/or reflection and sketch each function.

1. $f(x) = x^3 - 1$

1.__________________

____

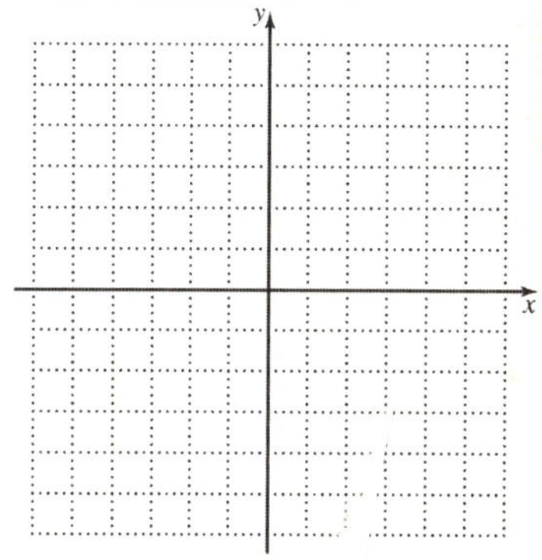

2. $g(x) = -|x+1|$

2.__________________

____

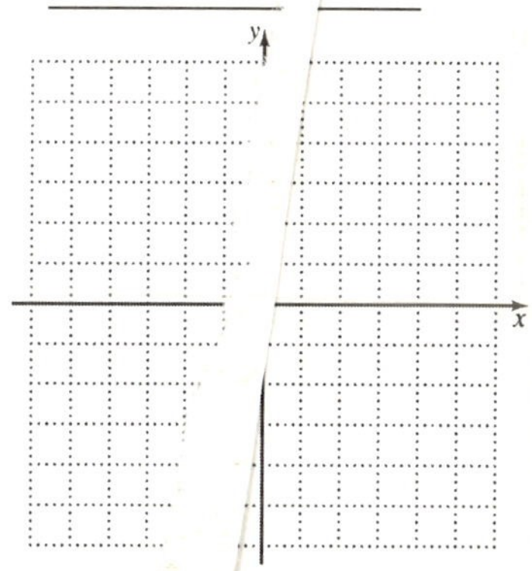

3. $h(x) = -x^2 + 3$

3.__________________

____

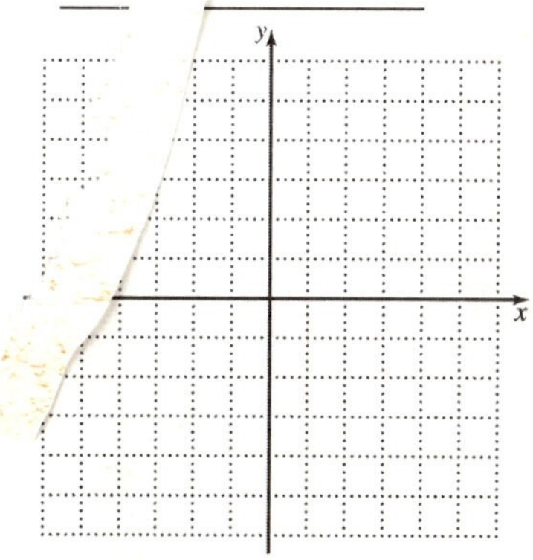

Name: Date:
Instructor: Section:

4. $f(x)=\sqrt{x+2}$

4. ____________________

____

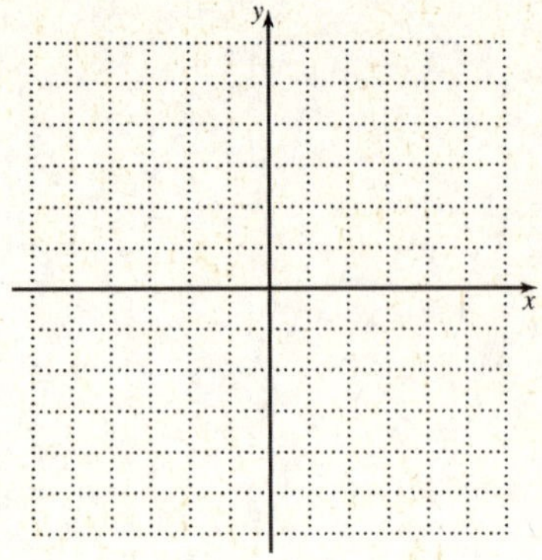

5. $g(x)=-x-3$

5. ____________________

____

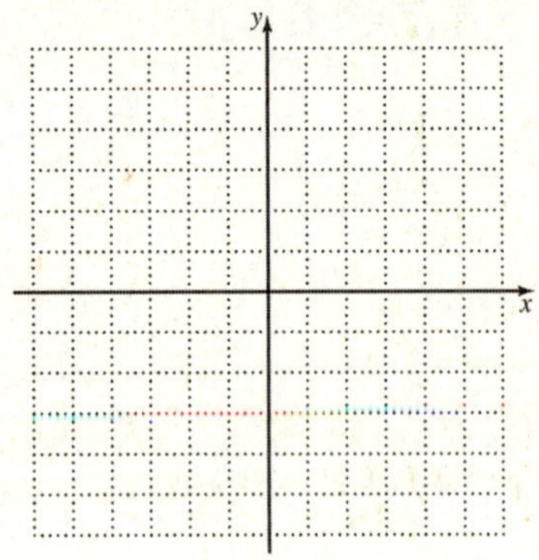

6. $h(x)=|x-3|-4$

6. ____________________

____

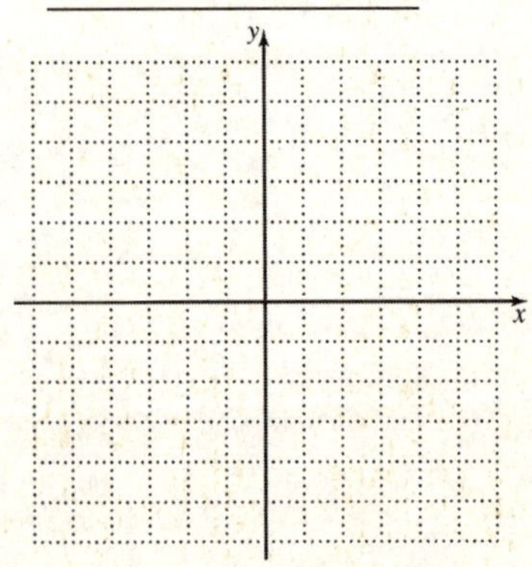

Name: Date:
Instructor: Section:

Chapter 10 COMPLEX NUMBERS AND QUADRATIC EQUATIONS

10.1 Complex Numbers

KEY VOCABULARY

Term	Definition	Example
Complex number written in standard form		

KEY PROPERTIES, PROCEDURES, OR STRATEGIES

Adding and Subtracting Complex Numbers

1.

2.

Multiplying Complex Numbers and Their Conjugates

1.

2.

Name: Date:
Instructor: Section:

GUIDED EXAMPLES

Perform the indicated operations and write the answers in the standard form $a + bi$.

a) $(12+5i)+(-16-7i)$

Solution

$(12+5i)+(-16-7i)$ — **Treat the imaginary unit *i* as you would a variable.**

$= \boxed{}$ — **Combine the real parts and the imaginary parts.**

$= \boxed{}$ — **Write in standard form for a complex number, $a + bi$.**

b) $(6-2i)(8+4i)$

Solution

$(6-2i)(8+4i)$

$= \boxed{}$ — **Use the FOIL method to multiply.**

$= \boxed{}$ — **Combine the imaginary parts and substitute $i^2 = -1$.**

$= \boxed{}$ — **Write in standard form.**

c) $\dfrac{-6+5i}{2i}$

Solution

$\dfrac{-6+5i}{2i} = \boxed{}$ — **Multiply by the special form of $1 = \dfrac{i}{i}$.**

$= \boxed{}$ — **Distribute *i* in the numerator and multiply the denominators.**

$= \boxed{}$ — **Substitute $i^2 = -1$ and simplify.**

$= \boxed{}$ — **Rewrite the numerator with the real part first. Write the answer in standard form $a + bi$.**

Name: Date:
Instructor: Section:

Chapter 10 COMPLEX NUMBERS AND QUADRATIC EQUATIONS

10.2 The Square Root Property, Completing the Square, and Quadratic Functions

KEY VOCABULARY

Term	Definition	Example
Quadratic equation		
Quadratic function		

KEY PROPERTIES, PROCEDURES, OR STRATEGIES

Zero Factor Property

In the Language of Math	In Your Own Words

Creating the Missing Term of a Perfect Square Trinomial

1.

2.

To Complete the Square on $ax^2 + bx + c = 0$

1.

2.

3.

4.

5.

Name: Date:
Instructor: Section:

GUIDED EXAMPLES

1. Solve the quadratic equation by factoring using the Zero Factor Property.

$5x^2 - 11x + 6 = 0$

Solution

$5x^2 - 11x + 6 = 0$

= [ ] **Factor the trinomial.**

= [ ] **Apply the Zero Factor Property and set each factor equal to 0.**

= [ ] **Solve each linear equation for *x*.**

Solution Set: [ ] **Check:**

2. Solve the quadratic equation by completing the square.

$x^2 - 5x = -8$

Solution

$x^2 - 5x = -8$ **$a = 1$ and the variable terms are isolated.**

$x^2 - 5x +$ [ ] $= -8 +$ [ ] **Complete the square.**

$(\quad)^2 =$ [ ] **Factor the perfect square trinomial as a binomial squared.**

[ ] or [ ] **Apply the Square Root Property.**

[ ] or [ ] **Simplify the square root and solve for *x*.**

[ ] or [ ] **Check:**

Solution Set: [ ]

Name: Date:
Instructor: Section:

PRACTICE PROBLEMS

Solve using the Zero Factor Property.

1. $(4x+1)(3x-7)=0$ **1.**______________

2. $10x^2+24x-18=0$ **2.**______________

3. $12x^2+11x=5$ **3.**______________

Solve using the Square Root Property.

4. $x^2=\dfrac{36}{81}$ **4.**______________

5. $x^2=-100$ **5.**______________

6. $4x^2+25=0$ **6.**______________

7. $(x-7)^2=9$ **7.**______________

8. $(3x-1)^2=25$ **8.**______________

Name: Date:
Instructor: Section:

9. $(4x-12)^2=-16$

9.________________

Find the missing term to make a perfect square trinomial and factor as a binomial squared.

10. $x^2-10x+\underline{\ ?\ }$

10.________________

11. $x^2+\frac{7}{3}x+\underline{\ ?\ }$

11.________________

Solve by completing the square.

12. $x^2-3x-40=0$

12.________________

13. $x^2+6x+10=0$

13.________________

14. $x^2-12x+22=8$

14.________________

15. $6x^2+18x+15=0$

15.________________

Name: Date:
Instructor: Section:

Chapter 10 COMPLEX NUMBERS AND QUADRATIC EQUATIONS

10.3 Quadratic Equations: The Quadratic Formula

KEY VOCABULARY

Term	Definition	Example
Discriminant		

KEY PROPERTIES, PROCEDURES, OR STRATEGIES

To Solve a Quadratic Equation Using the Quadratic Formula

1.

2.

3.

4.

5.

To Choose the Method for Solving any Quadratic Equation

1.

2.

3.

4.

Name: Date:
Instructor: Section:

GUIDED EXAMPLE

Solve $2x^2 + 11x - 9 = 0$ using the Quadratic Formula.

Solution

$2x^2 + 11x - 9 = 0$	**The equation is in standard form.**
$a =$ ☐ $b =$ ☐ $c =$ ☐	**Identify *a*, *b*, and *c*.**
☐	**Write the Quadratic Formula.**
☐	**Substitute values into the formula.**
☐	**Simplify the radicand.**
☐	**State the two solutions.**
Solution set: ☐	**Check:**

NOTES

Name: Date:
Instructor: Section:

PRACTICE PROBLEMS

Solve using the Quadratic Formula.

1. $2x^2 + x - 21 = 0$ **1.**____________

2. $3x^2 + 4x = 7$ **2.**____________

3. $2x^2 + 2x + 13 = 0$ **3.**____________

4. $6x^2 + 3x = 0$ **4.**____________

5. $3x^2 = 5x + 4$ **5.**____________

Use the discriminant to determine the number and type of solutions for each quadratic equation.

6. $10x^2 + 4x - 5 = 0$ **6.**____________

7. $5x^2 - 8x + 12 = 0$ **7.**____________

Name: Date:
Instructor: Section:

Solve by the method of your choice.

8. $4x^2 - 3x - 5 = 0$ **8.**____________

9. $8x^2 + 5x = 0$ **9.**____________

10. $(x+4)(x-7) = 12$ **10.**____________

11. $6x^2 + 8x + 3 = 0$ **11.**____________

12. $5x^2 + 8x - 3 = 0$ **12.**____________

Solve the equations involving rational expressions.

13. $\frac{5}{x} = \frac{x-3}{x+3}$ **13.**____________

14. $\frac{2x}{x+1} + \frac{3}{x} = \frac{x^2-3}{x^2-x}$ **14.**____________

Name: Date:
Instructor: Section:

Chapter 10 COMPLEX NUMBERS AND QUADRATIC EQUATIONS

10.4 Equations Quadratic in Form and Applications

GUIDED EXAMPLE

Solve $3(x-4)^2-(x-4)-10=0$ using substitution.

Solution

Let $u=(x-4)$, then $u^2=$ [ ]

Use substitution to rewrite the original equation in the new variable u.

$3(x-4)^2-(x-4)-10=0$

[ ] **Substitute $u=(x-4)$. The new equation is a quadratic equation in u.**

[ ] **Factor the LHS.**

[ ] or [ ] **Set each factor equal to 0 and solve for u.**

[ ] or [ ]

[ ] or [ ] **Resubstitute $(x-4)$ in for u and solve for x.**

[ ] or [ ]

Solution set: [ ] **Check:**

Name: Date:
Instructor: Section:

GUIDED EXAMPLE

Solve $2x^{-2}+8x^{-1}-10=0$ using substitution.

Solution

Let $u=$ ☐ then $u^2=$ ☐

Use substitution to rewrite the original equation in the new variable u.

$2x^{-2}+8x^{-1}-10=0$

☐ **Substitute $u=$ ☐**

The new equation is a quadratic equation in u.

☐ **Factor the LHS.**

☐ or ☐ **Set each factor equal to 0 and solve for u.**

☐ or ☐

☐ or ☐ **Resubstitute ☐ in for u and solve for x.**

☐ or ☐ **Apply the definition of negative exponents.**

☐ or ☐ **Clear fractions and note that the value $x = 0$ is excluded.**

☐ or ☐ **Solve for x.**

Solution set: ☐ **Check:**

Name: Date:
Instructor: Section:

PRACTICE PROBLEMS

Solve the following nonlinear equations.

1. $6x^4 + 17x^2 - 3 = 0$ **1.**________________

2. $x^4 - 11x^2 + 30 = 0$ **2.**________________

3. $12x^4 + 2x^2 = 14$ **3.**________________

4. $11(x+4)^2 + 21(x+4) - 2 = 0$ **4.**________________

5. $2x^{-2} - 9x^{-1} - 18 = 0$ **5.**________________

Name: Date:
Instructor: Section:

6. $x + x^{\frac{1}{2}} - 42 = 0$ **6.**______________

(Hint: Let $u = x^{\frac{1}{2}}$. Check for extraneous solutions)

Solve each formula for the indicated variable. Assume all variables represent positive real numbers.

7. $x = \dfrac{3a^2b^2}{2}$ for a **7.**______________

8. $y = a\sqrt{b^2 - 2c}$ for c **8.**______________

Solve the problem by utilizing the 4P's: Prepare, Plan, Process, and Ponder.

9. Tom and Huck have agreed to paint a wooden fence around their neighbor's yard. Working alone, it would take Tom 2 hours longer to do the job than it would take Huck. Working together, the boys can do the job in 5 hours. How long would it have taken each of them to do the job alone? Round your answers to the nearest tenth of an hour. **9.**______________

Name: Date:
Instructor: Section:

Chapter 10 COMPLEX NUMBERS AND QUADRATIC EQUATIONS

10.5 Quadratic and Rational Inequalities

KEY VOCABULARY

Term	Definition	Example
Quadratic inequality		

KEY PROPERTIES, PROCEDURES, OR STRATEGIES

To Solve a Quadratic Inequality

1.

2.

3.

4.

To Solve Rational Inequalities

1.

2.

3.

4.

Name: Date:
Instructor: Section:

GUIDED EXAMPLE

Solve $x^2+2x-24<0$. Graph the solutions on the real line and state the answer using interval notation.

Solution

The RHS is already zero, so factor the LHS.

$x^2+2x-24<0$

$\left(\quad \right)\left(\quad \right)<0$ **Use this inequality for the sign analysis.**

Find the *boundary values* by setting the quadratic *equal* to 0 and solving for x.

$\left(\quad \right)\left(\quad \right)=0$ **Set the quadratic equal to 0.**

☐ or ☐ **Apply the Zero Factor Property.**

☐ or ☐ **Solve for x. Use these values on the real line to create the test regions.**

Choose a test point within each region (other than the boundary values) to determine the sign of the quadratic in order to find which region(s) satisfy the inequality.

Region A	Region B	Region C
Test point: $x =$ ☐	Test point: $x =$ ☐	Test point: $x =$ ☐

Are the boundary values solutions to the inequality? ☐

Graph the solutions on the real line.

Solution Set: ☐

Name: Date:
Instructor: Section:

PRACTICE PROBLEMS

Solve. Graph solutions on the real number line and state the solutions using interval notation.

1. $(x-3)(x+4)<0$

1.________________

2. $x^2-4x-12\geq 0$

2.________________

3. $x^2-8x\leq 0$

3.________________

4. $2x^2-15x+18>0$

4.________________

5. $x^2-11x\geq -30$

5.________________

6. $4x^2-3x<0$

6.________________

7. $x^2-64\leq 0$

7.________________

Name: Date:
Instructor: Section:

8. $25x^2 > 36$

8.________________

9. $\frac{1}{x-5} > 0$

9.________________

10. $\frac{x+2}{x-7} \geq 0$

10.________________

11. $\frac{x-3}{x+9} \leq 0$

11.________________

12. $\frac{x}{x+4} \geq 0$

12.________________

13. $\frac{2x+5}{x-1} < 3$

13.________________

Name: Date:
Instructor: Section:

Chapter 11 CONIC SECTIONS

11.1 The Parabola

KEY PROPERTIES, PROCEDURES, OR STRATEGIES

Graphing Quadratic Equations in the Form $\begin{matrix} y = ax^2 + bx + c \\ x = ay^2 + by + c \end{matrix}$ **with** $a \neq 0$

$y = ax^2 + bx + c$	$x = ay^2 + by + c$

Graphing Quadratic Equations in the Form $\begin{matrix} y = a(x-h)^2 + k \\ x = a(y-k)^2 + h \end{matrix}$ **with** $a \neq 0$

$y = a(x-h)^2 + k$	$x = a(y-k)^2 + h$

General Form for Conic Sections

Name: Date:
Instructor: Section:

GUIDED EXAMPLE

Determine if the parabola opens up/down or left/right, the axis of symmetry, the vertex, and the intercepts. Then graph the parabola given by the quadratic equation. Label at least three points. $x = y^2 - 3y + 2$

Solution

Identify $a =$ ☐ $b =$ ☐ $c =$ ☐

Orientation Because the equation is quadratic in ☐ and $a =$ ☐

The parabola opens ☐

Axis of Symmetry

$\dfrac{-b}{2a} =$ ☐ $=$ ☐ so the axis of symmetry is the line ☐

Vertex

Substitute the value of $\dfrac{-b}{2a}$ into the original equation.

$$x = \left(\quad \right)^2 - 3\left(\quad \right) + 2$$

$x =$ ☐ The vertex is ☐

Intercepts

The x-intercept is $(c, 0) =$ ☐

Find the y-intercepts by setting $x = 0$ and solving for y.

$0 = y^2 - 3y + 2$

☐

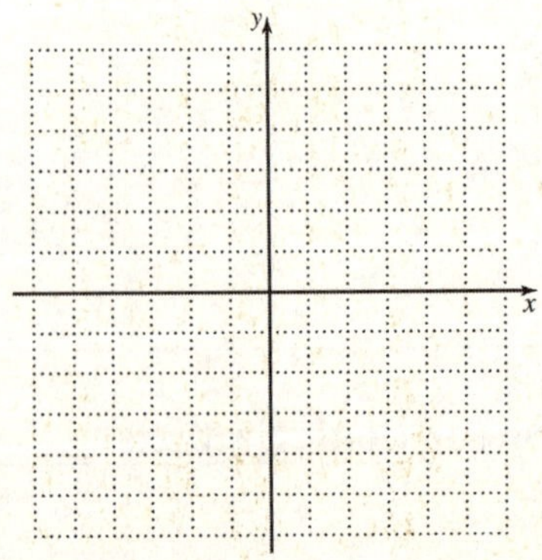

The y-intercepts are ☐ and ☐

Name: Date:
Instructor: Section:

PRACTICE PROBLEMS

Determine a) if each parabola opens up/down or left/right, b) the axis of symmetry, c) the vertex, and d) the intercepts. Then graph each parabola given by the quadratic equation. Label at least three points.

1. $y = -2x^2 + 4$

1a.________________

b.________________

c.________________

d.________________

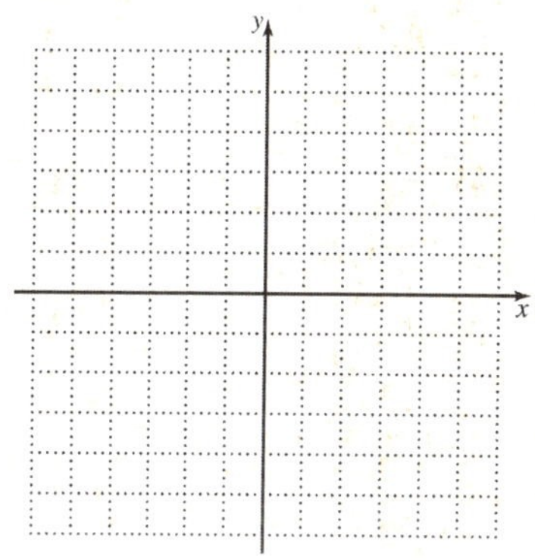

2. $y = x^2 + 4x - 1$

2a.________________

b.________________

c.________________

d.________________

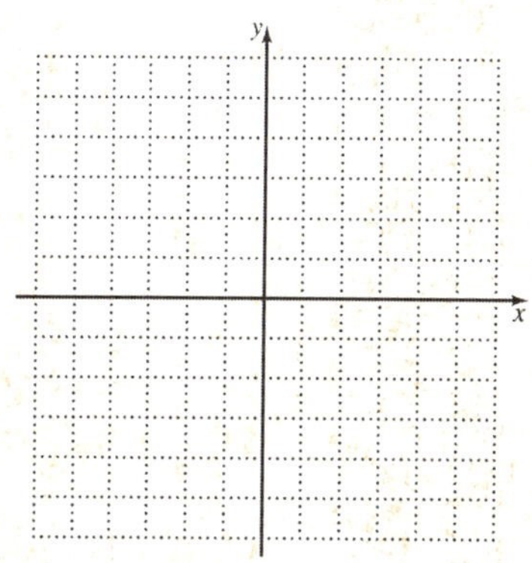

Name: Date:
Instructor: Section:

3. $x = y^2 - 3$

3a.________________

b.________________

c.________________

d.________________

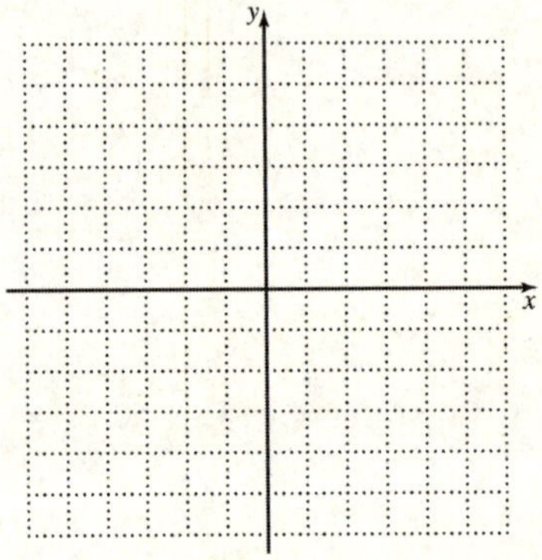

4. $x = -y^2 - 2y + 5$

4a.________________

b.________________

c.________________

d.________________

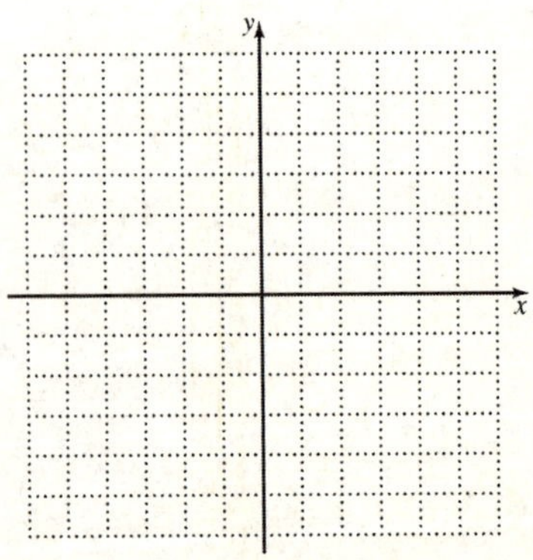

Name: Date:

Instructor: Section:

Graph the following quadratic equations.

5. $y=(x-1)^2+4$

5.

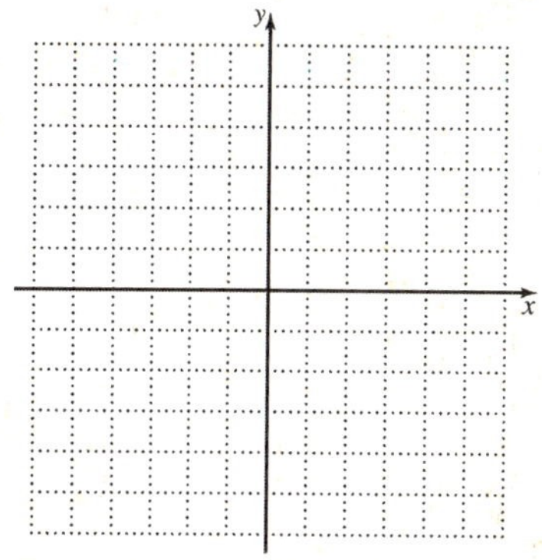

6. $y=-(x+4)^2-3$

6.

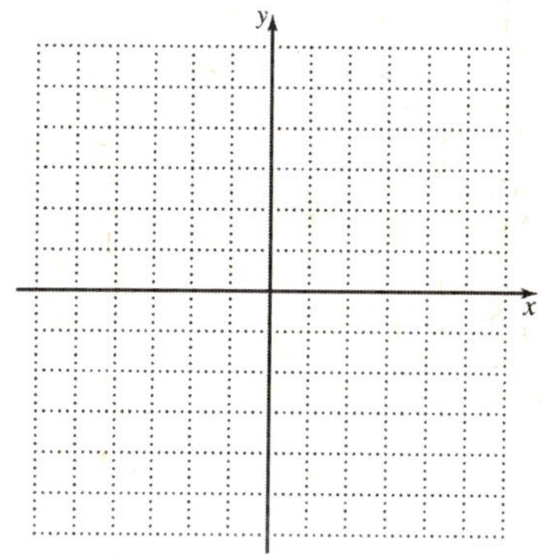

7. $x=(y+2)^2$

7.

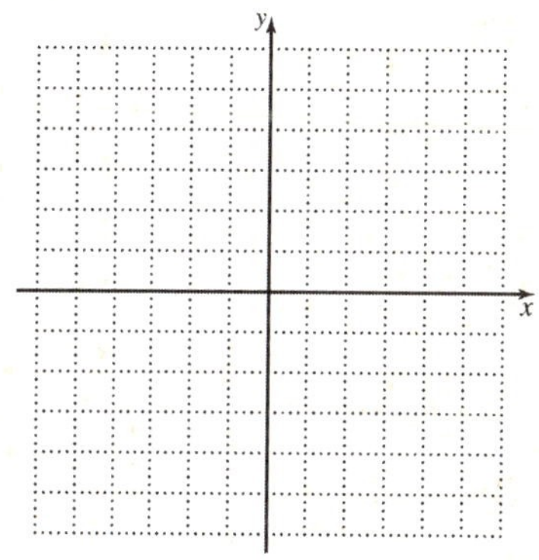

8. $x=-2(y-1)^2+3$

8.

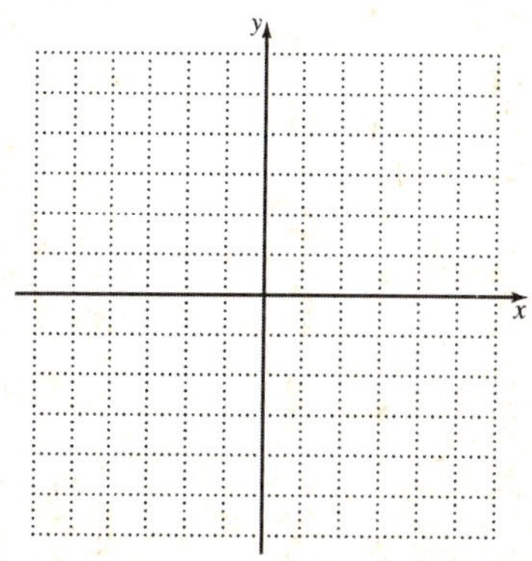

Name: Date:
Instructor: Section:

Use the process of completing the square to rewrite each quadratic equation in the form $y=a(x-h)^2+k$ *or* $x=a(y-k)^2+h$. *State a) the vertex, b) the axis of symmetry, and c) whether the parabolas open up/down or left/right. Do not graph.*

9. $y=x^2-12x+41$

9.______________

a)______________

b)______________

c)______________

10. $y=\frac{1}{2}x^2-3x+\frac{1}{2}$

10.______________

a)______________

b)______________

c)______________

11. $x=-3y^2+6y+2$

11.______________

a)______________

b)______________

c)______________

12. $x=2y^2+16y+31$

12.______________

a)______________

b)______________

c)______________

Name: Date:
Instructor: Section:

Chapter 11 CONIC SECTIONS

11.2 The Circle and Ellipse

KEY VOCABULARY

Term	Definition	Example
Circle		
Ellipse		

KEY PROPERTIES, PROCEDURES, OR STRATEGIES

Standard Form for the Equation of a Circle

To Graph a Circle in Standard Form

1.

2.

3.

General Form for Conic Sections

Name: Date:
Instructor: Section:

KEY PROPERTIES, PROCEDURES, OR STRATEGIES

Equation of an Ellipse Centered at the Origin

To Graph an Ellipse Centered at the Origin

1.

2.

3.

4.

Equation of an Ellipse with Center $C\ (h, k)$

GUIDED EXAMPLE

Graph the ellipse. Label the endpoints of the axes. $\frac{(x-4)^2}{9}+\frac{(y+1)^2}{25}=1$

Solution

Identify the center $C\ (h, k) =$ ______

$a^2 =$ ______ and $b^2 =$ ______

therefore $a =$ ______ and $b =$ ______

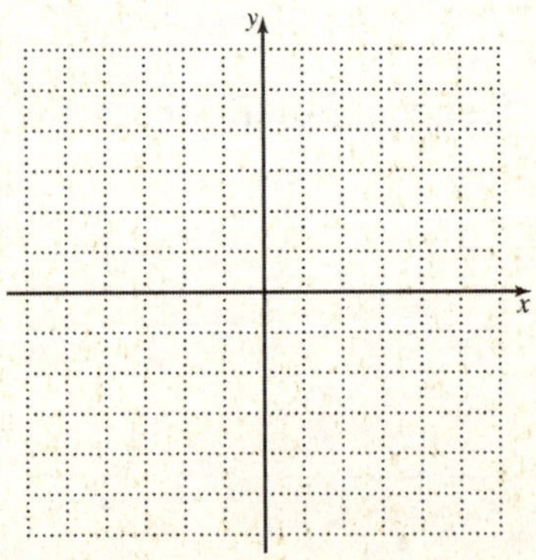

Use the center and the values of a and b to find the endpoints of each axis.

Name: Date:
Instructor: Section:

PRACTICE PROBLEMS

Sketch the graphs of the circles given by the following equations.

1. $x^2+(y+4)^2=4$

1.

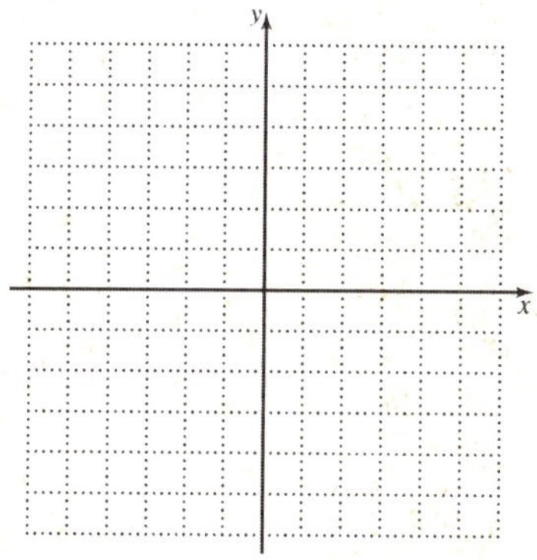

2. $(x+2)^2+(y-1)^2=16$

2.

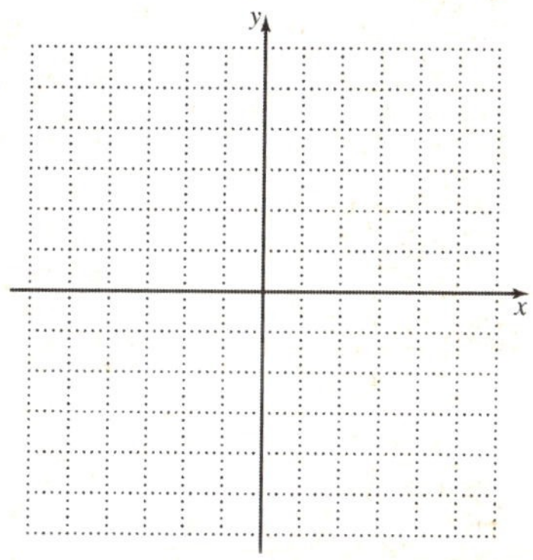

Write the equations of the circles in standard form, satisfying the given conditions.

3. Center $(-4,2)$; $r=\sqrt{2}$

3.________________

4. Center $(8,-1)$; $r=3\sqrt{6}$

4.________________

Name: Date:
Instructor: Section:

Use the process of completing the square to rewrite the equation of each circle in standard form. State the radius and the center of each circle.

5. $x^2 + y^2 + 10x - 4y + 13 = 0$

5.________________

____

6. $x^2 + y^2 - 12x + 6y + 20 = 0$

6.________________

____

Graph each ellipse. Label the intercepts.

7. $\frac{x^2}{16} + \frac{y^2}{25} = 1$

7.

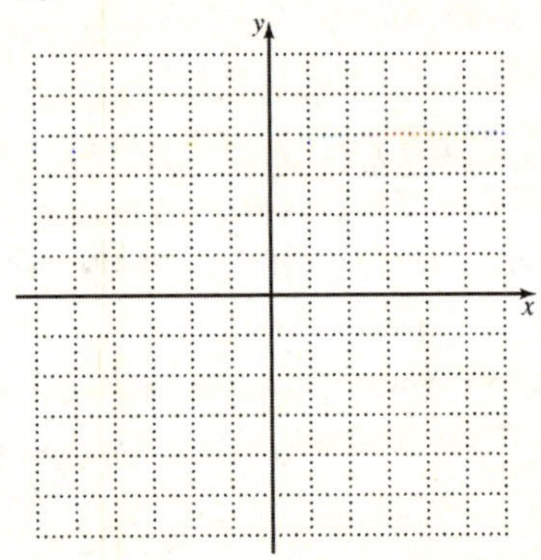

8. $4x^2 + y^2 = 36$

8.

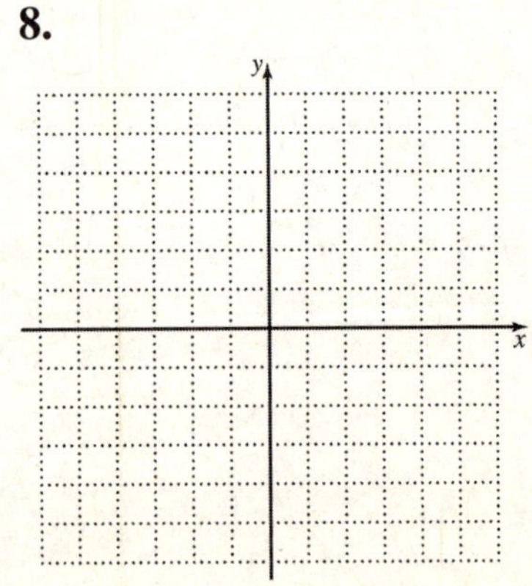

Name: Date:
Instructor: Section:

Graph each ellipse. Label the endpoints of the axes.

9. $\dfrac{(x-2)^2}{16}+\dfrac{(y+1)^2}{4}=1$

9.

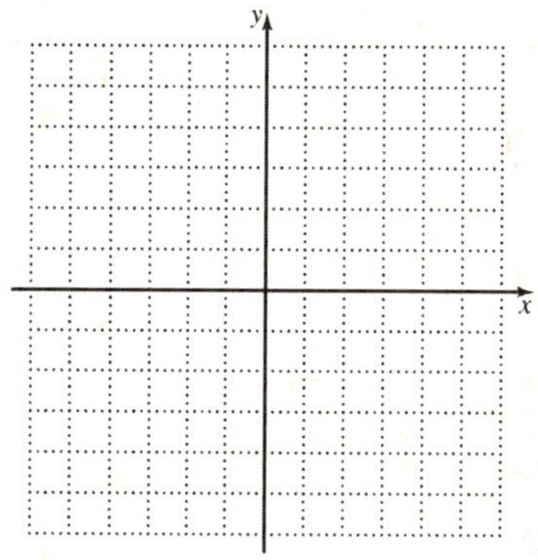

10. $\dfrac{(x+3)^2}{9}+\dfrac{y^2}{36}=1$

10.

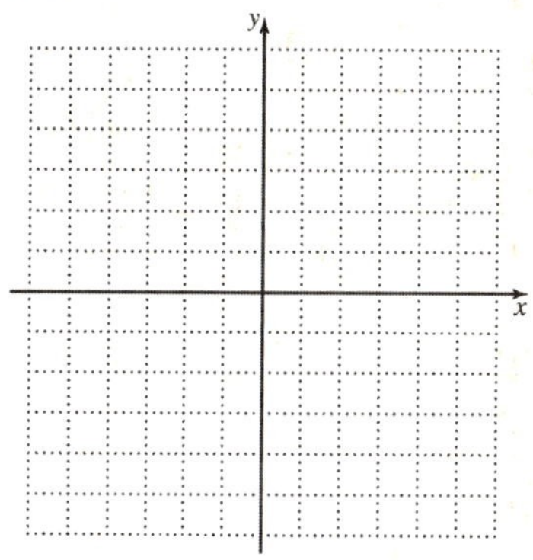

Write the equation of each ellipse in standard form, satisfying the following conditions.

11. The center is $(0,0)$. The x-intercepts are $(-7,0)$ and $(7,0)$. The y-intercepts are $(0,-9)$ and $(0,9)$.

11.________________

12. The center is $(4,-1)$. $a = 8$, $b = 10$.

12.________________

Name: Date:
Instructor: Section:

Identify each conic section as a parabola, a circle, or an ellipse, based on the general form $Ax^2+By^2+Cx+Dy+E=0$.

13. $y^2+2x-30=0$ **13.**______________

14. $5x^2+3y^2-8=0$ **14.**______________

15. $4x^2+4y^2-18=0$ **15.**______________

Name: Date:
Instructor: Section:

Chapter 11 CONIC SECTIONS

11.3 The Hyperbola

KEY VOCABULARY

Term	Definition	Example
Hyperbola		

KEY PROPERTIES, PROCEDURES, OR STRATEGIES

Equations of Hyperbolas Centered at the Origin

$\frac{x^2}{a^2} - \frac{y^2}{b^2} = 1$ with $a, b > 0$	$\frac{y^2}{b^2} - \frac{x^2}{a^2} = 1$ with $a, b > 0$

To Graph a Hyperbola Centered at the Origin

1.

2.

3.

4.

5.

Name: Date:
Instructor: Section:

Equations of a Hyperbola with the Center $C\ (h, k)$

$\frac{(x-h)^2}{a^2} - \frac{(y-k)^2}{b^2} = 1$	$\frac{(y-k)^2}{b^2} - \frac{(x-h)^2}{a^2} = 1$

General Form for Conic Sections

$Ax^2 + By^2 + Cx + Dy + E = 0$, with A, B, C, D, E, and E real numbers and A and B both not equal to zero

Type of Conic Section	Number of Squared Terms	Further Restrictions

GUIDED EXAMPLE

Write the equation of the hyperbola with the conditions $C\ (0,0)$, vertices $(0,-5)$ and $(0,5)$, and two additional points $(-6,0)$ and $(6,0)$.

Solution

We want the form of a hyperbola centered at the origin that opens up and down: $\frac{y^2}{b^2} - \frac{x^2}{a^2} = 1$

Vertices $(0,-5)$ and $(0,5)$ mean that $b =$ ☐ and so $b^2 =$ ☐

Two additional points $(-6,0)$ and $(6,0)$ mean that $a =$ ☐ and so $a^2 =$ ☐

So the equation becomes ☐

Name: Date:
Instructor: Section:

PRACTICE PROBLEMS

Graph each hyperbola. Label the vertices.

1. $\dfrac{x^2}{4}-\dfrac{y^2}{4}=1$

1.

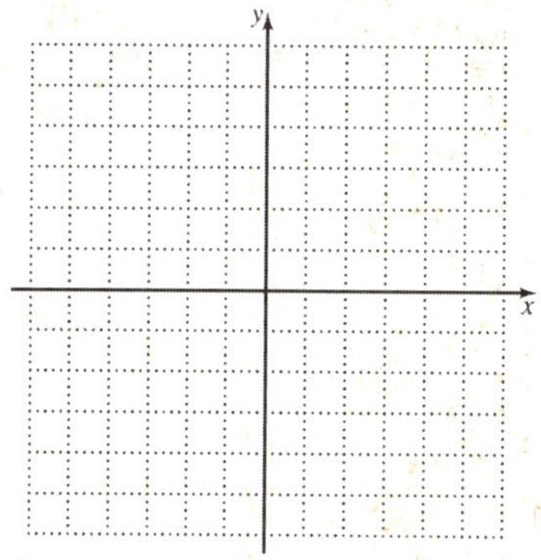

2. $\dfrac{y^2}{9}-\dfrac{x^2}{16}=1$

2.

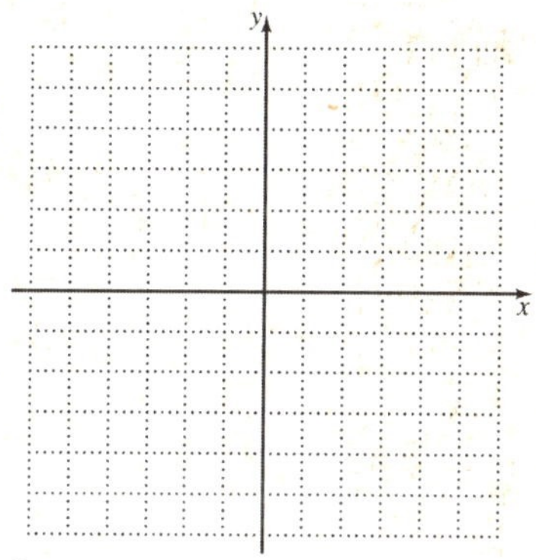

3. $x^2-y^2=1$

3.

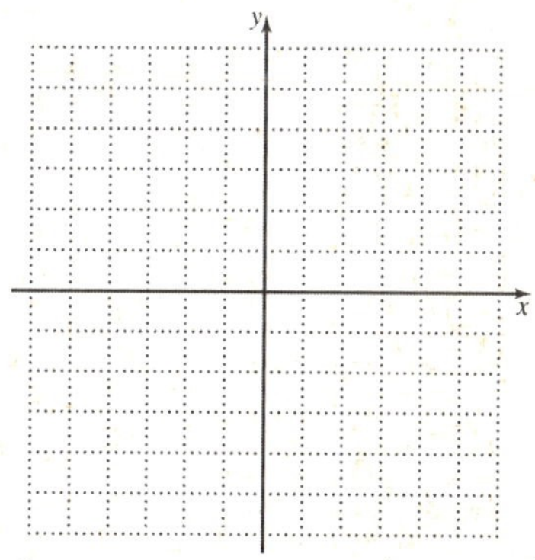

4. $\dfrac{(y-1)^2}{4}-\dfrac{(x+2)^2}{9}=1$

4.

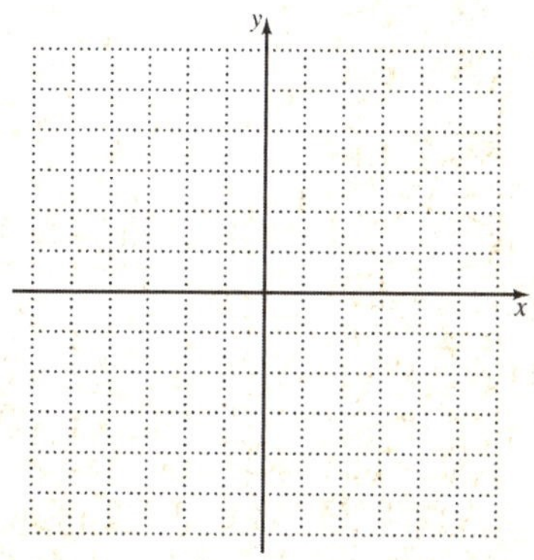

Name: Date:
Instructor: Section:

5. $\frac{(x+3)^2}{4}-\frac{(y-2)^2}{4}=1$

5.

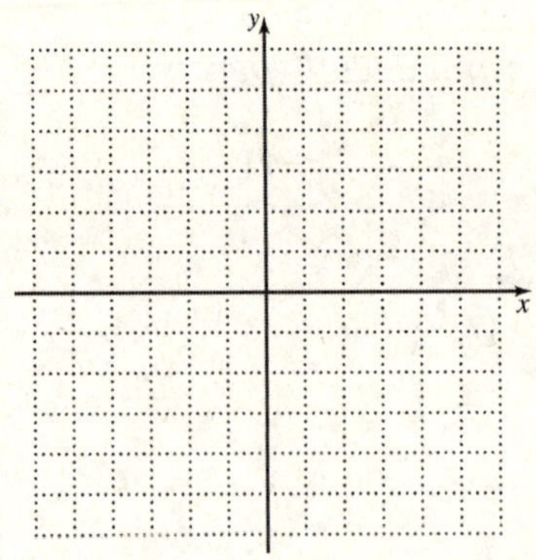

Write the equations of the hyperbolas satisfying the following conditions.

6. $C\ (0,0)$, vertices $(0,-8)$ and $(0,8)$, two additional points $(-2,0)$ and $(2,0)$.

6.________________

7. $C\ (-4,5)$, opening left and right with $a = 3$, $b = 6$

7.________________

Identify each equation as either a line or a type of conic section (parabola, circle, ellipse, or hyperbola) based on the general form $Ax^2+By^2+Cx+Dy+E=0$.

8. $16x^2+y^2-64=0$

8.________________

9. $-x^2+y^2-9=0$

9.________________

10. $3x^2-5y+4=0$

10.________________

Name: Date:
Instructor: Section:

Chapter 11 CONIC SECTIONS

11.4 Systems of Nonlinear Equations and Inequalities

KEY VOCABULARY

Term	Definition	Example
System of nonlinear equations		
Nonlinear polynomial inequality		
System of nonlinear inequalities		

NOTES

Name: Date:

Instructor: Section:

GUIDED EXAMPLE

Solve the system of nonlinear equations for real solutions.

$$\begin{cases} 4x^2 + y^2 = 16 \\ y = x^2 - 4 \end{cases}$$

Solution

First, identify the conic sections in the system.

$\begin{cases} 4x^2 + y^2 = 16 \\ y = x^2 - 4 \end{cases}$ $\leftarrow$ Two squared terms, coefficients have same sign, not equal:

$\leftarrow$ One squared term:

Make a quick sketch to determine the number of possible solutions.

How many solutions are possible?

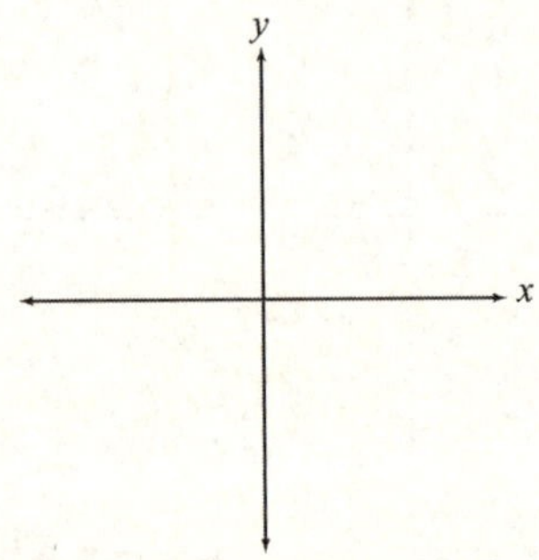

Which solution method appears to be the best choice? The second equation is already solved for *y*, so the method may be the best choice.

$$4x^2 + y^2 = 16$$

$$4x^2 + \left(\qquad \right)^2 = 16$$ **Substitute for *y*.**

Simplify the LHS.

The equation is quadratic in form; factor and use the Zero Factor Property.

Solve for *x*.

Substitute each x value back into $y = x^2 - 4$ to find its corresponding y-value.

Solution Set:

Check each solution point in the original equations.

Name: Date:
Instructor: Section:

PRACTICE PROBLEMS

Solve each system of nonlinear equations using the Graphing Method.

1. $\begin{cases} x^2 + y^2 = 9 \\ x - y = 3 \end{cases}$

1.________________

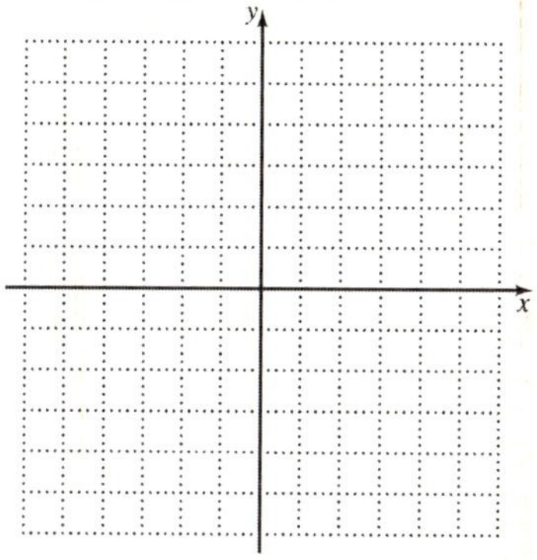

2. $\begin{cases} y = -x^2 + 4 \\ y = -x - 2 \end{cases}$

2.________________

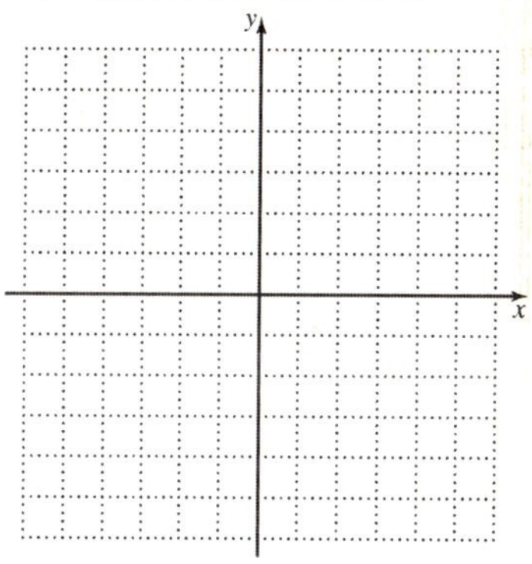

Solve each system of nonlinear equations for real solutions.

3. $\begin{cases} (x+1)^2 + (y-2)^2 = 25 \\ 5(x+1)^2 + (y-2)^2 = 125 \end{cases}$

3.________________

4. $\begin{cases} x^2 + 4y^2 = 16 \\ x^2 + y^2 = 4 \end{cases}$

4.________________

Name: Date:
Instructor: Section:

5. $\begin{cases} x = y^2 + 6 \\ x = -y^2 + y - 8 \end{cases}$

5.________________

Solve each nonlinear inequality by graphing.

6. $y \geq x^2 - 3$

6.

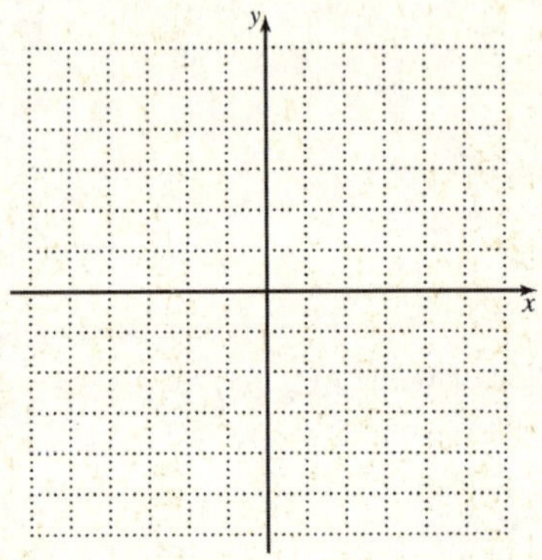

7. $(x-1)^2 + y^2 < 9$

7.

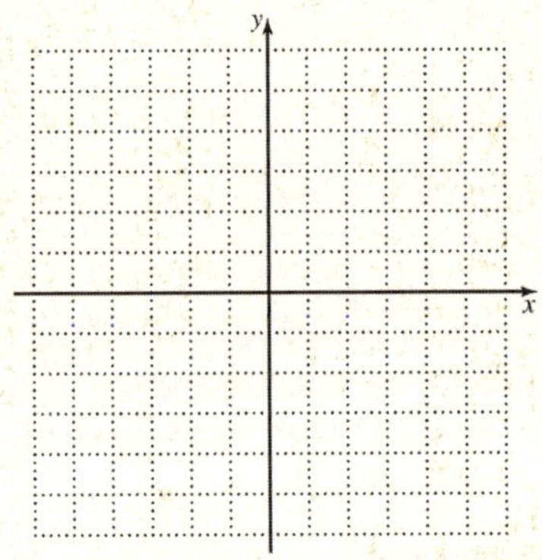

Solve each system of nonlinear inequalities by graphing.

8. $\begin{cases} \dfrac{x^2}{16} + \dfrac{(y+2)^2}{9} \leq 1 \\ y \geq x^2 + 4x + 2 \end{cases}$

8.

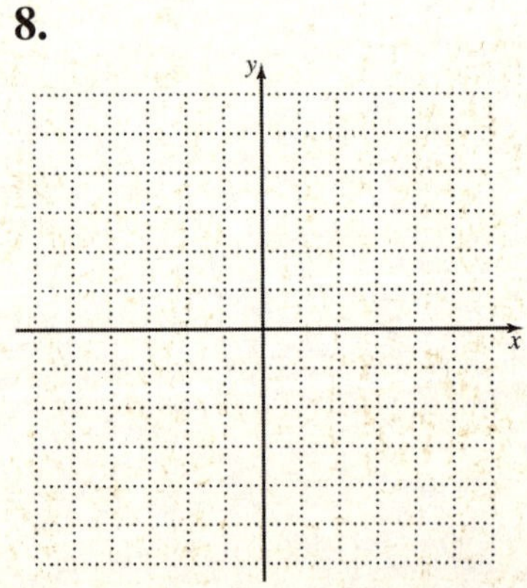

Name: Date:
Instructor: Section:

Chapter 12 INVERSE, EXPONENTIAL, AND LOGARITHMIC FUNCTIONS

12.1 Inverse and One-to-One Functions

KEY VOCABULARY

Term	Definition	Example
Relation		
Function		
One-to-one function		
Inverse function		

KEY PROPERTIES, PROCEDURES, OR STRATEGIES

Using the Horizontal Line Test to Determine One-to-One Functions

To Find the Inverse Function of a One-to-One Function

1.

2.

3.

4.

Name: Date:
Instructor: Section:

GUIDED EXAMPLE

Find the inverse function of $f(x)=\sqrt[3]{x+5}$.

Solution

$f(x)=\sqrt[3]{x+5}$

	Replace $f(x)$ with y.
	Interchange x and y.
	Solve for y.
	Replace y with $f^{-1}(x)$.

NOTES

Name: Date:
Instructor: Section:

PRACTICE PROBLEMS

Determine if each function is one-to-one. Explain your answers.

1. $f(x) = -4x + 5$ **1.**________________

____

2. $g(x) = 2x^2 - 7$ **2.**________________

____

3. $h(x) = |3x + 6|$ **3.**________________

____

4. $g(x) = -2x^3$ **4.**________________

____

5. $f(x) = \frac{x+3}{5}$ **5.**________________

____

Determine if the functions are inverses of each other.

6. $f(x) = 7x,\ g(x) = \frac{x}{7}$ **6.**________________

7. $f(x) = x + 9,\ g(x) = x - 9$ **7.**________________

Name: Date:
Instructor: Section:

8. $f(x) = 3x + 4,\ g(x) = 2 - 3x$ 8._______________

9. $f(x) = x^7 + 5,\ g(x) = \sqrt[7]{x - 5}$ 9._______________

Find the inverse function of each one-to-one function. Then, verify that $(f \circ f^{-1})(x) = x$, $(f^{-1} \circ f)(x) = x$, $D_f = R_{f^{-1}}$ *and* $R_f = D_{f^{-1}}$.

10. $f(x) = 6x + 5$ 10._______________

11. $g(x) = \sqrt{x - 9}$ 11._______________

12. $h(x) = (x + 8)^3$ 12._______________

13. $f(x) = \dfrac{10}{x}$ 13._______________

Name: Date:
Instructor: Section:

Chapter 12 INVERSE, EXPONENTIAL, AND LOGARITHMIC FUNCTIONS

12.2 Exponential Functions

KEY VOCABULARY

Term	Definition	Example
Exponential function		
Natural exponential function		

KEY PROPERTIES, PROCEDURES, OR STRATEGIES

Graphs of Exponential Functions

Exponential Properties

1.

2.

Name: Date:
Instructor: Section:

GUIDED EXAMPLES

Solve the following exponential equations.

a) $3^x = 243$

Solution

$3^x = 243$

To use the exponential property $b^x = b^y$ if and only if $x = y$, the bases must be the same. Rewrite each side with the same base.

$3^x =$ ☐

$x =$ ☐

$b^x = b^y$ if and only if $x = y$.

Solution set: ☐

Check:

b) $2^{4x-5} = \frac{1}{128}$

Solution

$2^{4x-5} = \frac{1}{128}$

$2^{4x-5} =$ ☐

Rewrite each side with the same base.

$4x - 5 =$ ☐

$b^x = b^y$ if and only if $x = y$.

☐

Solve for x.

Solution set: ☐

Check:

NOTES

Name: Date:
Instructor: Section:

PRACTICE PROBLEMS

Graph each exponential function. Label the intercept and the asymptote.

1. $f(x) = 4^{x+1}$

1.

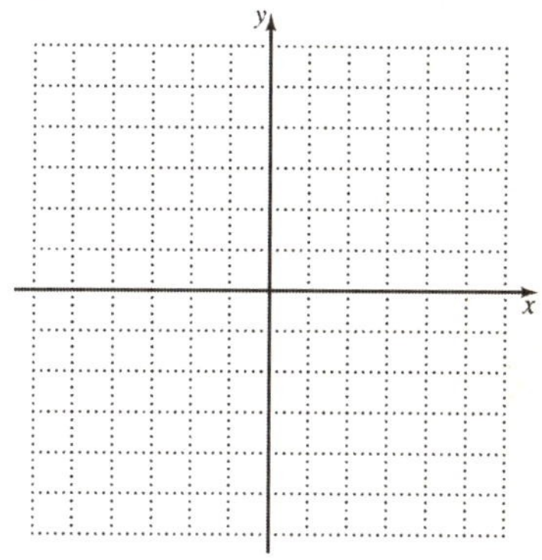

2. $g(x) = 4^{x} + 1$

2.

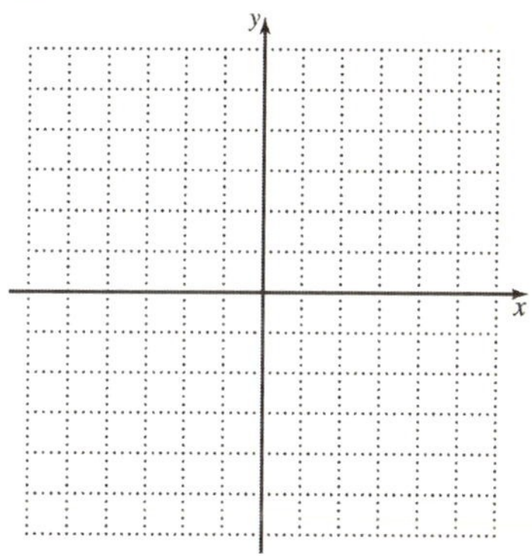

3. $h(x) = \left(\frac{1}{4}\right)^{x}$

3.

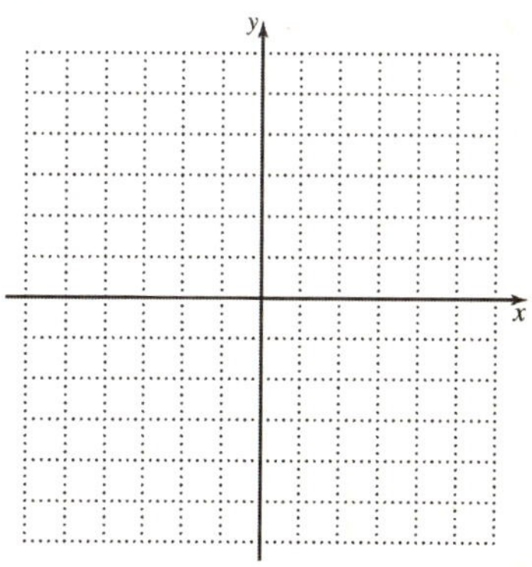

Name: Date:
Instructor: Section:

Using a calculator, evaluate the following natural exponential expressions for the indicated values of the independent variable. Round the answers to three decimal places.

4. $f(x) = -7e^{4.1x}$; $x = -1, 0, 3$

4.________________

____

____

5. $f(x) = 2e^{\frac{x}{4}}$; $x = -3, 0, 1$

5.________________

____

____

Solve the following exponential equations.

6. $2^x = 64$

6.________________

7. $27^x = 3$

7.________________

8. $3^{5x-1} = 81$

8.________________

9. $5^{7x+4} = \frac{1}{125}$

9.________________

Name: Date:
Instructor: Section:

Chapter 12 INVERSE, EXPONENTIAL, AND LOGARITHMIC FUNCTIONS

12.3 Logarithmic Functions

KEY VOCABULARY

Term	Definition	Example
Logarithm with base *b* of *x*		
Logarithmic function with base *b*		
Common logarithmic function		
Natural logarithmic function		

KEY PROPERTIES, PROCEDURES, OR STRATEGIES

The Graph of the Logarithmic Function

Name: Date:
Instructor: Section:

GUIDED EXAMPLES

Solve the following logarithmic equations by converting to exponential form.

a) $\log_4(x) = 5$

Solution

$\log_4(x) = 5$

$\log_4(x) = 5$ if and only if [ ] **Convert to exponential form.**

$x =$ [ ] **Solve for *x*.**

b) $\log_y\left(\frac{1}{2}\right) = -\frac{1}{3}$

Solution

$\log_y\left(\frac{1}{2}\right) = -\frac{1}{3}$

$\log_y\left(\frac{1}{2}\right) = -\frac{1}{3}$ if and only if [ ] **Convert to exponential form.**

[ ] **Apply exponent rules to simplify.**

[ ] **Solve for *y*.**

NOTES

Name: Date:
Instructor: Section:

PRACTICE PROBLEMS

Write in exponential form.

1. $\log_4(256) = 4$ **1.**______________

2. $\log_8\left(\frac{1}{64}\right) = -2$ **2.**______________

3. $\frac{1}{3} = \log_{64} 4$ **3.**______________

Write each in logarithmic form.

4. $4^5 = 1024$ **4.**______________

5. $\frac{1}{27} = 3^{-3}$ **5.**______________

6. $81^{\frac{1}{2}} = 9$ **6.**______________

Evaluate each logarithm by using the definition.

7. $\log_7 343$ **7.**______________

8. $\log_2\left(\frac{1}{32}\right)$ **8.**______________

9. $\log_{512} 8$ **9.**______________

Name: Date:
Instructor: Section:

Graph the logarithmic function. State the domain and range using interval notation.

10. $f(x) = \log_6(x)$ **10.**________________

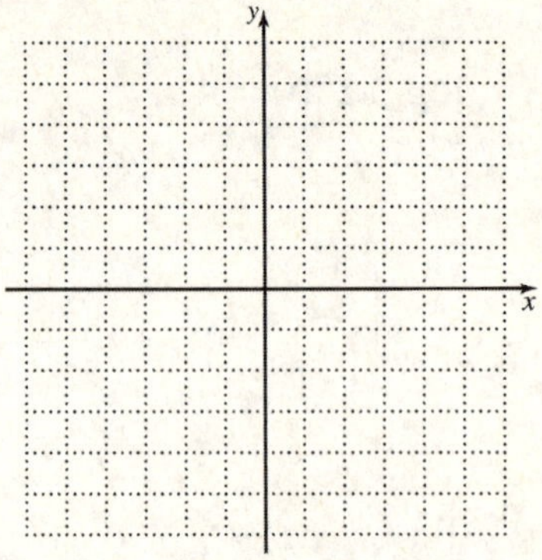

Use a calculator to evaluate the common and natural logarithms. Round the answers to three decimal places.

11. $\log(0.002)$ **11.**________________

12. $\ln\left(\sqrt{3}+8\right)$ **12.**________________

Solve the following logarithmic equations by converting to exponential form.

13. $\log_{216}(x) = \frac{1}{3}$ **13.**________________

14. $\log_{10,000}(10) = z$ **14.**________________

15. $\log_a\left(\frac{1}{81}\right) = -4$ **15.**________________

Name: Date:
Instructor: Section:

Chapter 12 INVERSE, EXPONENTIAL, AND LOGARITHMIC FUNCTIONS

12.4 Properties of Logarithms

KEY PROPERTIES, PROCEDURES, OR STRATEGIES

Properties of Logarithms

Evaluation Properties:

1.

2.

Inverse Property:

3.

4.

Operational Properties of Logarithms:

5.

6.

7.

NOTES

Name: Date:
Instructor: Section:

GUIDED EXAMPLES

Use the properties of logarithms to write each expression as a sum, difference, and/or multiple of logarithms.

a) $\log_5\left(\frac{4x^7}{y^3}\right)$

Solution

$\log_5\left(\frac{4x^7}{y^3}\right)$

$\log_5\left(\frac{4x^7}{y^3}\right) = \square$ **Apply the quotient rule to get a difference of logs.**

$= \square$ **Apply the product rule to get a sum of logs.**

$= \square$ **Apply the power rule to pull down the exponent from the argument.**

$= \square$

b) $\ln\left(y\sqrt{6x^3}\right)$

Solution

$\ln\left(y\sqrt{6x^3}\right)$

$\ln\left(y\sqrt{6x^3}\right) = \square$ **Apply the product rule to get a sum of logs.**

$= \square$ **Write the square root as an exponential expression.**

$= \square$ **Apply the power rule to pull down the exponent from the argument.**

$= \square$ **Apply the product rule to get a sum of logs.**

$= \square$ **Apply the power rule to pull down the exponent from the argument.**

$= \square$

Name: Date:
Instructor: Section:

PRACTICE PROBLEMS

Use a logarithmic property to simplify each of the following expressions.

1. $\log_5(5)$ **1.**______________

2. $\log_9(1)$ **2.**______________

3. $8^{\log_8(3)}$ **3.**______________

Use the properties of logarithms to write each expression as a sum, difference, and/or multiple of logarithms, if possible.

4. $\log_2(a^7)$ **4.**______________

5. $\log_6\left(\frac{8}{y}\right)$ **5.**______________

6. $\ln\left(\sqrt{3b}\right)$ **6.**______________

7. $\log_4\left(\frac{7x}{9}\right)$ **7.**______________

8. $\log_5[3(s+t)]$ **8.**______________

Name: Date:
Instructor: Section:

Write each expression as a logarithm of a single quantity.

9. $6\ln(x)+\ln(y)$ **9.**________________

10. $\log_2(18)-\log_2(3)$ **10.**________________

11. $4\log(9)+4\log(x)$ **11.**________________

12. $\frac{1}{5}\log_b(y)+6\log_b(z-4)$ **12.**________________

13. $\ln(32)-4\ln(2)+\ln(5)$ **13.**________________

Given $\log_b(2)\approx 0.36$, $\log_b(3)\approx 0.56$, *and* $\log_b(5)\approx 0.83$, *use the properties of logarithms to approximate the following to two decimal places.*

14. $\log_b(6)$ **14.**________________

15. $\log_b\left(\frac{75}{8}\right)$ **15.**________________

Name: Date:
Instructor: Section:

Chapter 12 INVERSE, EXPONENTIAL, AND LOGARITHMIC FUNCTIONS

12.5 Logarithmic and Exponential Equations and Applications

KEY PROPERTIES, PROCEDURES, OR STRATEGIES

The Change-of-Base Formula

In the Language of Math	In Your Own Words

The One-to-One Logarithmic Property

Strategy for Solving Exponential Equations

1.

2.

To Solve Logarithmic Equations

1.

2.

3.

Name: Date:
Instructor: Section:

GUIDED EXAMPLE

Solve the exponential equation $4^{3x-1} = 3^{x+2}$. Round the answer to two decimal places, if necessary.

Solution

$4^{3x-1} = 3^{x+2}$

	Take the natural logarithm of both sides.
	Apply the power rule of logarithms.
	Distribute on each side.
	Gather variables on the LHS and constants on the RHS.
	Factor out the *x*.
	Solve for *x* and round if necessary.

Check:

NOTES

Name: Date:
Instructor: Section:

PRACTICE PROBLEMS

Using natural logarithms and the change-of-base formula, evaluate the following logarithmic expressions. Round each answer to four decimal places.

1. $\log_6(14)$ **1.**________________

2. $\log_{\frac{1}{5}}\left(\sqrt{7}\right)$ **2.**________________

Using common logarithms and the change-of-base formula, evaluate the following logarithmic expressions. Round each answer to four decimal places.

3. $\log_{\sqrt{5}}(2{,}146)$ **3.**________________

4. $\log_{2.4}\left(\frac{7}{8}\right)$ **4.**________________

Solve the following exponential equations. Round each answer to two decimal places, if necessary.

5. $4^{2x+5}=762$ **5.**________________

6. $e^{-4x}=5.1$ **6.**________________

7. $6^{3x+1}=10^{4x-5}$ **7.**________________

Name: Date:
Instructor: Section:

Solve each logarithmic equation. Check for extraneous solutions.

8. $\log_2(x+5)=9$ **8.**______________

9. $\log(8x+3)=\log(-4x-5)$ **9.**______________

10. $\log(x)-\log(x+4)=\log(3)$ **10.**______________

11. $\log_2(x^2-6x+16)=3$ **11.**______________

12. $\log(2x+8)-\log(4)=\log(x+1)+\log(x-1)$ **12.**______________

Solve the following application using the compound interest formula $A=P\left(1+\frac{r}{n}\right)^{nt}$, *where A is the balance in an account with an initial investment P, for t years, at an annual percentage rate r for n compoundings per year.*

13. Charlotte and Miranda each decided to invest \$10,000 for 5 years. Charlotte invested her money at an annual rate 4.35% compounded weekly and Miranda invested her money at 6.0% compounded monthly. Which woman made a better investment? **13.**______________

Name: Date:
Instructor: Section:

Appendix B RATIO AND PROPORTION

KEY VOCABULARY

Term	Definition	Example
Ratio		
Proportion		

KEY PROPERTIES, PROCEDURES, OR STRATEGIES

To Solve a Proportion

In the Language of Math	In Your Own Words

NOTES

Name: Date:
Instructor: Section:

GUIDED EXAMPLES

1. Write the following ratio in fraction form.
Five pints to 6 quarts

Solution
Convert the larger unit of measure, quarts, to the smaller unit of measure, pints. That is, 1 quart = 2 pints.

$$\frac{5\text{ pints}}{6\text{ quarts}} = \frac{5\text{ pints}}{6(2)\text{ pints}} = \frac{5\text{ pints}}{12\text{ pints}}$$

2. Solve the proportion $\frac{x}{115} = \frac{7}{35}$ for x.

Solution

$$\frac{x}{115} = \frac{7}{35}$$

	=		**The product of the extremes is equal to the product of the means.**
	=		
	=		**Solve for *x*.**
	=		

Name: Date:
Instructor: Section:

PRACTICE PROBLEMS

1. In an afternoon yoga class there are 10 women and 4 men. What is the ratio of women to the total number of people in the class? Write the ratio in three ways.

1.________________

2. In a carton of a dozen eggs three eggs are cracked and the rest are intact. What is the ratio of the number of eggs that are not cracked to the number of eggs that are cracked? Write the ratio as a fraction.

2.________________

Write the following ratios in fraction form.

3. Six months to four years

3.________________

4. Three yards to two feet

4.________________

5. Ninety minutes to five hours

5.________________

Solve each proportion.

6. $\frac{x}{15} = \frac{12}{45}$

6.________________

Name: Date:
Instructor: Section:

7. $\frac{52}{x} = \frac{56}{98}$ 7.________________

8. $\frac{27}{63} = \frac{x}{42}$ 8.________________

9. On a map's scale, 5 centimeters represents 25 miles. What will be the length on the map that would represent a distance of 200 miles? 9.________________

10. A model of a proposed office building is made to scale. One-half inch on the model represents ten feet. How long is the enclosed elevated walkway going to be if it is eight inches on the model? 10.________________

Chapter 1 REVIEW OF REAL NUMBERS AND PROBLEM SOLVING

1.1 Sets, Symbols, and Real Numbers

1. false, $3 \notin A$ 3. false, $100 \in A$ 5. $\left\{-95, -77, 0, 2, \sqrt{9}, \frac{14}{3}, 5, 8.25, 9.\overline{3}\right\}$

7. $\left\{-95, -77, 0, 2, \sqrt{9}, 5\right\}$ 9. $\left\{-8.252252225..., -\sqrt{17}, -\pi, \sqrt{7}\right\}$

11. $\{4, 5, 6, 7, ...\}$ 13. $\{\ \}$ or $\varnothing$ 15. false, $-17 < -8$

1.2 Review of Fractions and Decimals

1. $2 \cdot 2 \cdot 3 \cdot 3 \cdot 3 \cdot 5$ 3. $\frac{1}{9}$ 5. $\frac{5}{16}$ 7. $\frac{3}{5}$ 9. $\frac{95}{72}$ 11. $\frac{53}{18}$

13. 18.714 15. 1.76

1.3 Addition and Subtraction of Real Numbers

1. 17 3. 30 5. -24 7. $-\frac{5}{4}$ 9. 31 11. $\frac{59}{40}$

13. -17 15. -272.3

1.4 Multiplication, Division, and Properties of Real Numbers

1. $\frac{1}{12}$ 3. $12c+12d$ 5. $\frac{1}{5}+\frac{1}{4}$ 7. $(7p+5q)+3r$

9. $a \cdot 8+(4+7)$ 11. 0 13. 18.4 15. -10

1.5 Exponents, Order of Operations, and Algebraic Expressions

1. 11.3 3. -47 5. -1 7. -288 9. $-0.9m-3$

11. $-\frac{5}{3}a+\frac{43}{5}b$ 13. $-17t-21$ 15. $-\frac{37}{15}d-\frac{1}{12}e$

1.6 Reading and Interpreting Graphs and the Coordinate Plane

1. 35 books 3. 3,000 people 5. 6% 7. \$45,000

1.7 Problem Solving: The 4P's Strategy

1. 3 nickels, 2 dimes, and 6 quarters 3. 101, 108, 115

5. 6 tomato plants, 4 lettuce plants, and 8 carrot plants 7. Amanda

9.

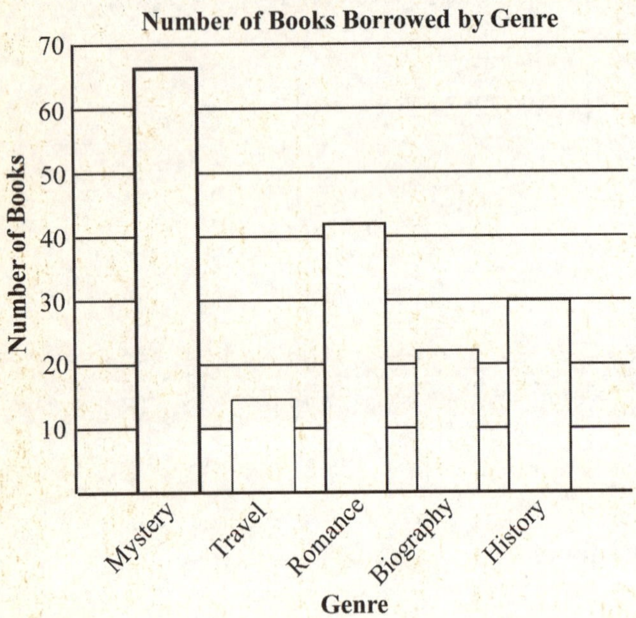

Chapter 2 EQUATIONS AND INEQUALITIES IN ONE VARIABLE WITH APPLICATIONS

2.1 Properties of Equality and Linear Equations

1. Yes 3. No 5. $\{24\}$ 7. $\{0\}$ 9. $\{10\}$ 11. $\{126\}$

13. $\{0\}$ 15. $\{6.5\}$

2.2 Linear Equations Involving Grouping Symbols and Fractions

1. $\{2\}$ 3. $\{-3\}$ 5. $\{7\}$ 7. $\{8\}$ 9. $\{x \mid x \in R\}$

11. $\{\ \}$ or $\varnothing$ 13. 65.2°, 24.8°

2.3 Literal Equations and Applications

1. $x = 35 - 5a$ 3. $A = \dfrac{D}{BC}$ 5. $K = \dfrac{4s - mc}{m}$ 7. $59°\text{F}$

9. \$65.00 11. five years 13. \$7,680.00 15. 15 min.

2.4 Linear Inequalities

1. $[7,\infty)$; 

3. $[-24,\infty)$; 

5. $(-\infty,4]$; 

7. $[130,\infty)$ 9. $(-\infty,-2]$

11. $\left[\frac{3}{2},\infty\right)$ 13. $(-\infty,\infty)$ 15. $(-\infty,\infty)$

2.5 Compound and Double Inequalities

1. $(-\infty,17)\cup[27,\infty)$; 

3. $(-\infty,-1]\cup(9,\infty)$; 

5. $(-7,32)$ 7. $(-\infty,2)\cup(3,\infty)$ 9. $\left(-\infty,-\frac{53}{3}\right)$ 11. $(22,27)$

13. $[2,29]$ 15. $(117,434]$

2.6 Absolute Value Equations and Inequalities

1. $\{-14,14\}$ 3. $\left\{\frac{4}{3},\frac{38}{3}\right\}$ 5. $\left\{-\frac{8}{9},2\right\}$ 7. $\{5\}$ 9. $\left\{-\frac{3}{5},-\frac{5}{9}\right\}$

11. $(-62,64)$; 

13. $\{\ \}$;

15. $(-\infty,-61)\cup(93,\infty)$; 

2.7 Additional Applications

1. $\frac{2}{3}$ hour, or 40 minutes 3. \$6200 at 6.5% and \$3800 at 8% 5. 3.6 pounds

Chapter 3 LINEAR EQUATIONS, SLOPE, INEQUALITIES, AND INTRODUCTION TO FUNCTIONS

3.1 Linear Equations in Two Variables

1. $5x-1=-20$ 3. yes   5. yes   7. $\frac{22}{3}$ 9. 7.5

11.

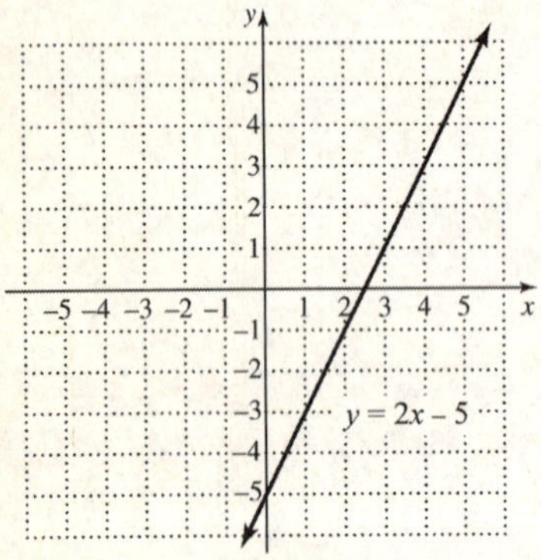

3.2 The Intercept Method

1.

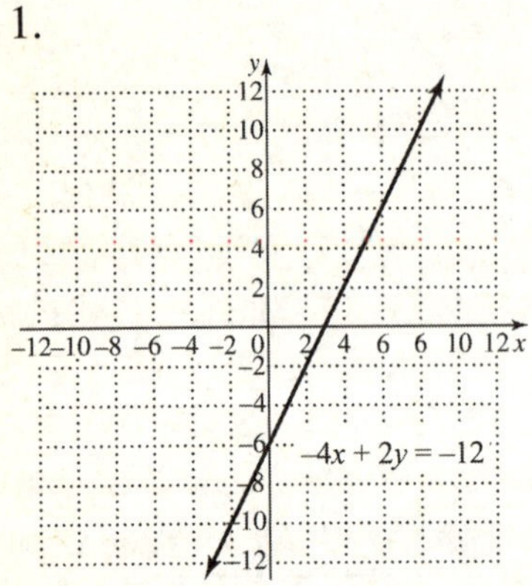

3.

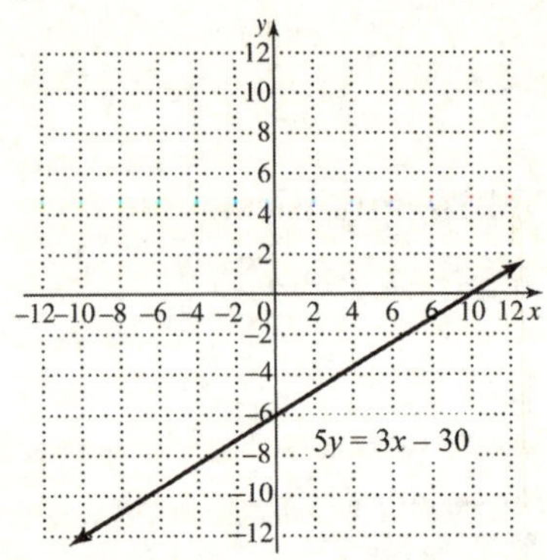

5.

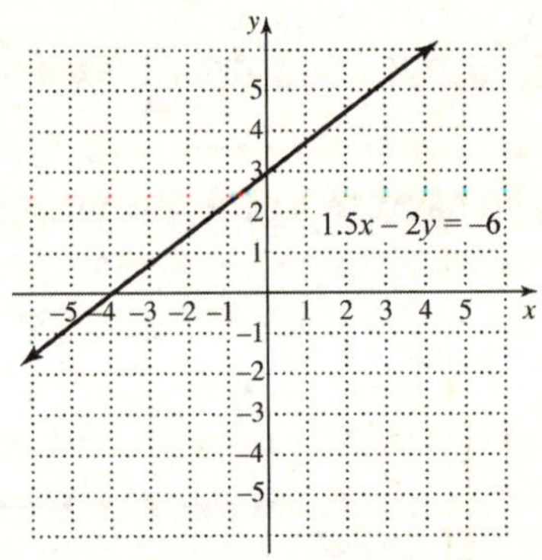

7. vertical   9. vertical

3.3 The Slope and *y*-Intercept Method

1. $\frac{3}{2}$ 3. $\frac{3}{4}$ 5. $-\frac{1}{12}$ 7. $m=-7;\ (0,1)$ 9. $m=-\frac{4}{5};\ (0,0)$

11. m is undefined; no y-intercept

13.

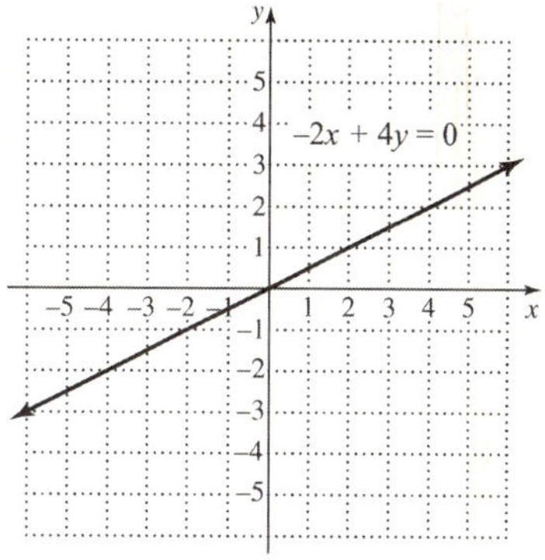

3.4 Determining the Equation of a Line, Parallel Lines, and Perpendicular Lines

1. $y = \frac{1}{4}x + 5$
3. $y = -\frac{5}{8}x - \frac{5}{8}$
5. $y = -\frac{5}{9}x - \frac{7}{3}$
7. perpendicular
9. neither
11. $y = -\frac{5}{3}x - \frac{17}{3}$
13. $x = -\frac{2}{3}$
15. $3x - 2y = -6$

3.5 Graphing Linear Inequalities in Two Variables

1.

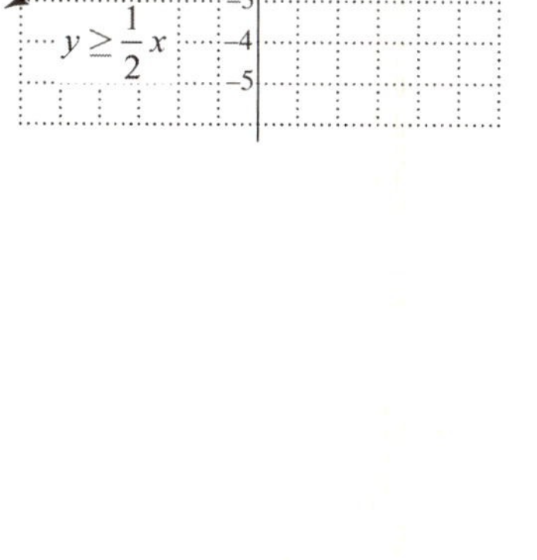

3.

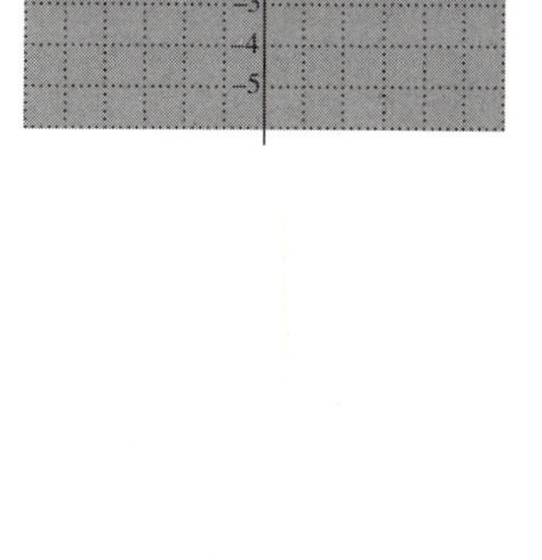

5.

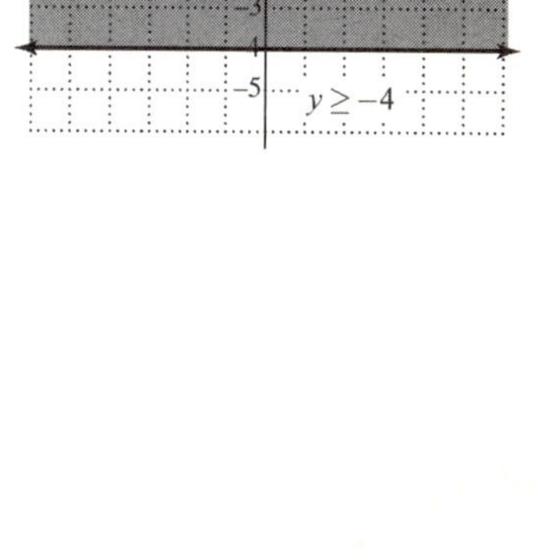

7.

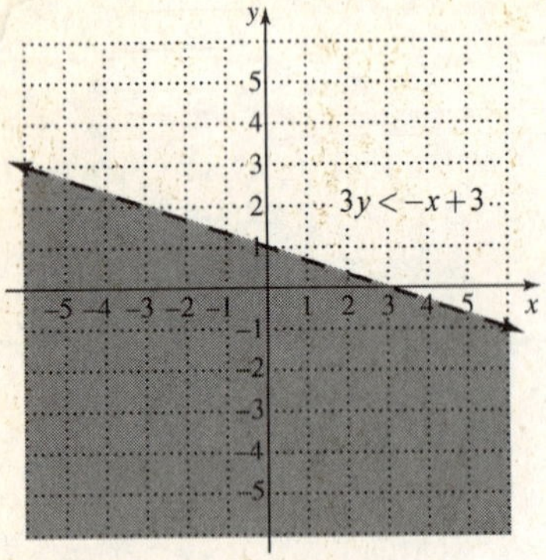

3.6 Introduction to Functions

1. relation 3. relation 5. $D = (-\infty, \infty); R = (-\infty, \infty)$

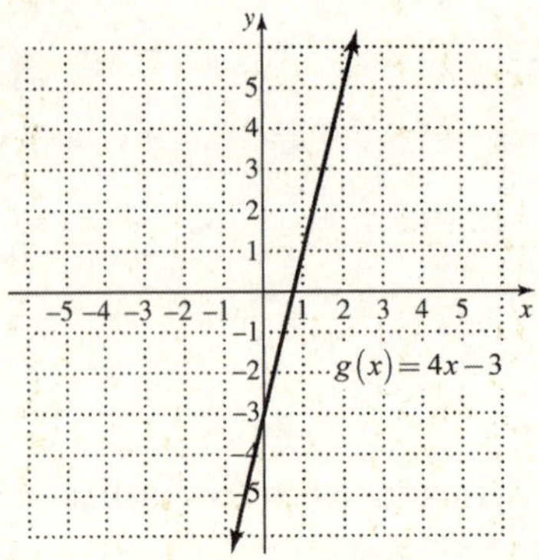

7. $(4, -23), (-8, 61)$ 9. $(5,6), \left(-4, \frac{15}{4}\right)$ 11. $(-1,14), (0.3, 7.89)$

13. $(18,14), (-11,15)$

Chapter 4 LINEAR SYSTEMS OF EQUATIONS AND INEQUALITIES

4.1 The Graphing Method

1. no 3. yes

5. no solution

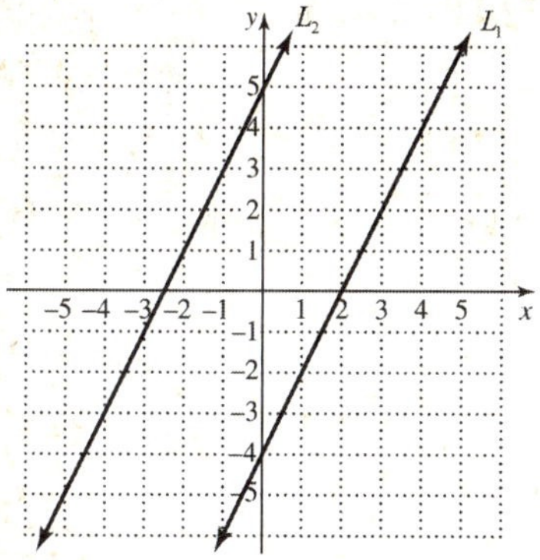

7. infinitely many solutions

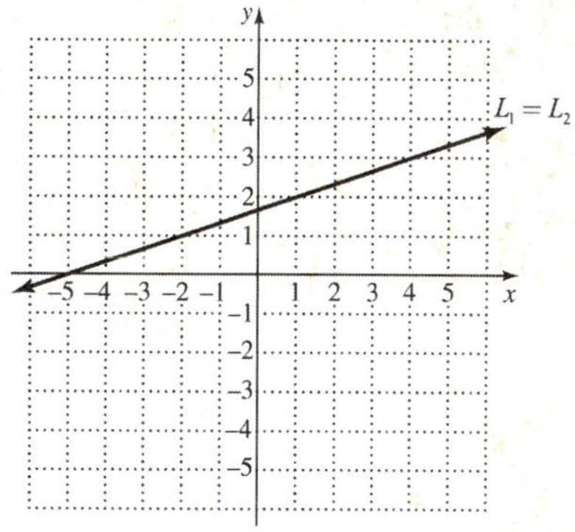

4.2 The Substitution Method

1. $\{(-2,0)\}$ 3. $\left\{\left(\frac{11}{4},-\frac{9}{4}\right)\right\}$ 5. $\left\{\left(-27,\frac{15}{2}\right)\right\}$ 7. $\{(-8,-2)\}$

9. $\left\{\left(-\frac{2}{3},\frac{1}{3}\right)\right\}$

4.3 The Elimination Method

1. $\{(1,5)\}$ 3. $\{(2,2)\}$ 5. $\{(1,2)\}$ 7. $\{(2,1)\}$ 9. $\{\ \}$

4.4 Systems of Linear Equations in Three Variables

1. $\{(-7,2,-2)\}$ 3. $\{(x, y, z)|3x+y-2z=2\}$ 5. $\left\{\left(-1,\frac{5}{3},\frac{1}{2}\right)\right\}$

4.5 Applications

1. $\{(-2,-1)\}$ 3. $\{(0,-1,-1)\}$ 5. 15 and $\frac{1}{5}$ 7. 56 and 58

4.6 Systems of Linear Inequalities

1.

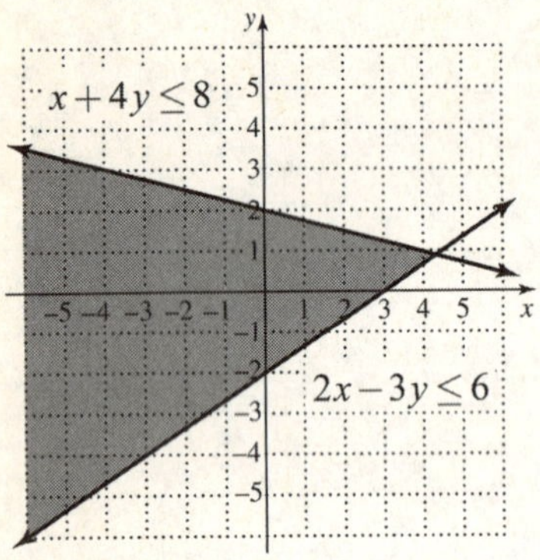

3.

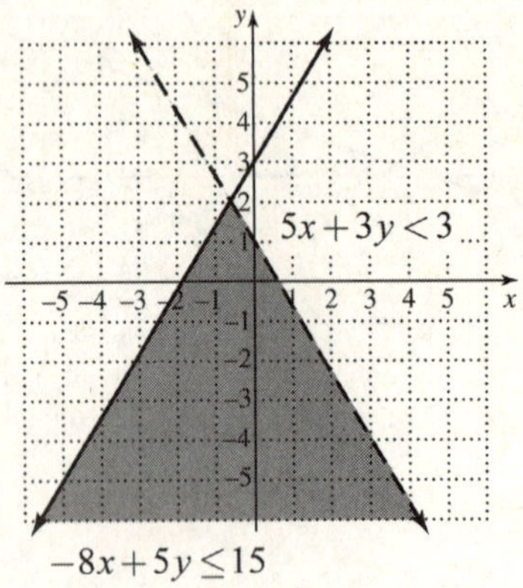

5.

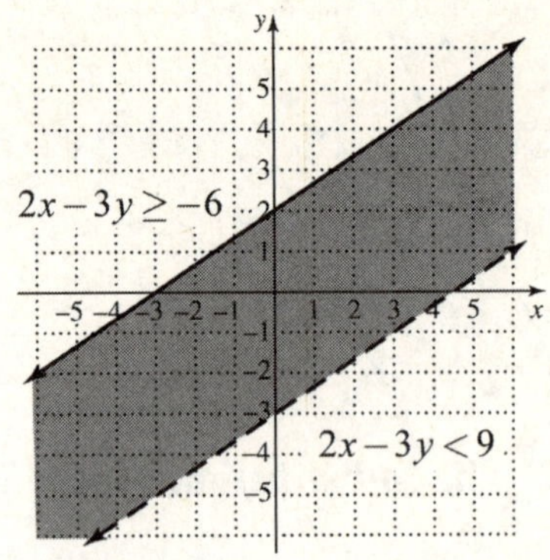

7.

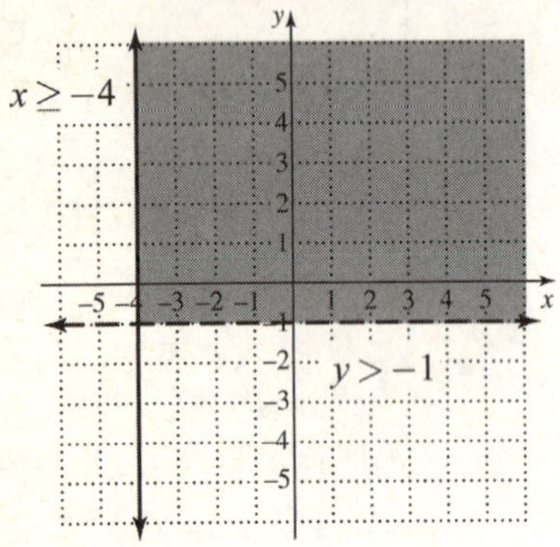

Chapter 5 EXPONENTS, POLYNOMIALS, AND APPLICATIONS

5.1 Integer Exponents

1. 8   3. x^2 5. d^{15} 7. $-27t^{15}u^6$ 9. z^5 11. $\dfrac{9b^{14}}{25c^8}$

13. $\dfrac{9n^3}{4m^4}$ 15. $\dfrac{-49r^{11}}{s^2}$

5.2 Scientific Notation and Applications

1. $4.56\left(10^6\right)$ 3. $5.6\left(10^{-7}\right)$ 5. 650,000,000   7. 0.0000156

9. 2,100,000,000,000; $2.1\left(10^{12}\right)$ 11. 300,000; $3\left(10^5\right)$ 13. 202.8; $2.028\left(10^2\right)$

15. 4,704,000 ft^2; $4.704(10^6)$ ft^2

5.3 Introduction to Polynomials and Polynomial Functions

1. 5   3. $7y^3-3y^2-3y+11$ 5. $\frac{1}{2}a^3+\frac{1}{3}a+\frac{4}{5}$

7. $-8x^4-8x^3+6x^2+2x+5$ 9. $7a^3+a+16$ 11. $\frac{1}{4}a^3-\frac{5}{6}ab+\frac{3}{4}b^2$

13. -332 15. 15.65

5.4 Multiplying Polynomials

1. $-8x^5y$ 3. $-\frac{8}{5}a^4b^8c^5$ 5. $24y^5+18y^4$ 7. $\frac{3}{10}r^2s^3-\frac{6}{5}r^4s^2$

9. $z^2+15z+54$ 11. $12a^2+14ab-10b^2$ 13. $-70a^3b^3+20a^2b^4-30ab^5$

15. $x^4+4x^3y-10x^2y^2+31xy^3-4y^4$

5.5 Special Products

1. $12a^2+ab-b^2$ 3. $-3a^4-9a^2b-2a^2b^2-6b^3$ 5. $x^2-6xz+9z^2$

7. $64x^4-48x^2y^2+9y^4$ 9. $49a^2-b^2$ 11. x^4-36y^2

13. $14a^2-22ab-12b^2$ 15. $49x^2-42xy+9y^2$

5.6 Dividing Polynomials

1. $4x+5$ 3. $\frac{1}{y^3}-\frac{1}{3xy}+\frac{4}{x^2}$ 5. $2x^2-8x+6$ 7. $4y^2-1+\frac{13}{2y+1}$

9. $8x^3-10x^2+11x-17+\frac{17}{x+1}$

Chapter 6 FACTORING POLYNOMIALS

6.1 The Greatest Common Factor and Factoring by Grouping

1. x^2 3. $4r^2(7r^2+3r)$ 5. $9(3-4x+7y)$ 7. $7z^3(4z^2+3z-8)$

9. $(x+4)(x-5)$ 11. $(3a+b)(2a-b)$ 13. $9(2x+1)(c-4d)$

15. $2x^2(x-4)(3x+y)$

6.2 Factoring Special Forms of Binomials

1. $(x+4)(x-4)$ 3. prime 5. $(6a+3b)(6a-3b)$

7. $(y+5)(y^2-5y+25)$ 9. $(2x-1)(4x^2+2x+1)$ 11. $7(x+2)(x-2)$

13. $8(x^3-2y^3)$ 15. $5(x^2+9)(x+3)(x-3)$

6.3 Factoring Trinomials of the Form $Ax^2 + Bx + C, A \neq 1$

1. $(2x+3)(x+1)$ 3. prime 5. $(2x-7)(4x-3)$ 7. $(3t+2)(8t-5)$

9. $x(3x+1)(x-7)$ 11. $3a(5a+1)(2a-3)$

6.4 Factoring Trinomials of the Form $Ax^2 + Bx + C$, with $A = 1$, and Perfect Square Trinomials

1. $(x+5)(x+3)$ 3. $(b-6)^2$ 5. $(x+5)^2$ 7. $(7x-y)^2$

9. $2z(z+4)(z-6)$ 11. $y^2(y-2)(y+1)$

6.5 The General Factoring Strategy

1. $5a^2(a-3)^2$ 3. $x^2z^3(z-4)(z^2+4z+16)$ 5. $(5x+2)(x+7)$

7. $8(3x-4)(3x-1)$ 9. $4x^3y(7x^2+30xy-8y^2)$ 11. $3x^4y^6z^5(2x^2+5z^3)$

13. prime

6.6 Applications Using the Zero Factor Property and Polynomial Functions

1. $\{-4,0\}$ 3. $\{-6,6\}$ 5. $\{-11,8\}$ 7. $\left\{-\frac{1}{4},\frac{5}{3}\right\}$ 9. $\left\{\frac{5}{8}\right\}$

11a. 1 b. 1 13. $L = 8$ meters, $W = 5$ meters

Chapter 7 RATIONAL EXPRESSIONS

7.1 Simplifying Rational Expressions

1. $\frac{7x^7}{8}$ 3. $\frac{y-4}{y+4}$ 5. $\frac{2}{x-4}$ 7. $-2-5y$ 9. $-x^2-3x-9$

11. $\frac{-y-5}{y-5}$

7.2 Multiplying and Dividing Rational Expressions

1. $\frac{6a}{7b}$ 3. 1 5. 1 7. $\frac{xy+4y^2}{9x}$ 9. $\frac{5d}{7c}$ 11. $\frac{16a^4b^4}{4a^4b^4c^4}$

7.3 Addition and Subtraction of Rational Expressions

1. $\frac{2a+3}{a+7}$ 3. $x-8$ 5. x^2-36 7. $\frac{2c^2+7c+21}{c(c+3)}$ 9. 1

11. $\frac{11}{x-3}$

7.4 Synthetic Division and Complex Fractions

1. $x-7$ 3. $x^2+x-6+\frac{31}{x+4}$ 5. $3x^2+3x-3$ 7. 2

9. $5x+y$

7.5 Solving Equations Containing Rational Expressions and Rational Functions

1. $D=\{t|t\neq -6\}$ 3. $D=\{x|x\neq -5 \text{ and } x\neq 3\}$ 5. $\{3\}$ 7. $\{-1\}$

9. equation, $\{-1,12\}$ 11. $T_1=\frac{P_1V_1T_1}{P_2V_2}$

7.6 Applications and Variations

1. $G=\frac{kab}{c}$ 3. 21 5. \$102.00

7. John drove at an average speed of 45 mph on his way to see Jane; he returned home at an average speed of 40 mph.

Chapter 8 ROOTS AND RADICALS

8.1 Simplifying, Adding, and Subtracting Square Roots

1. 10 3. $4\sqrt{3}$ 5. $8|x|$ 7. $|a-6|$ 9. $z^9\sqrt{z}$

11. $8m^7n^8\sqrt{3mn}$ 13. $-2\sqrt{5}+4\sqrt{7}$ 15. $18\sqrt{3}$

8.2 Multiplying and Dividing Square Roots

1. $5\sqrt{3}$ 3. $5x^2\sqrt{14x}$ 5. $3+\sqrt{15}$ 7. $-4\sqrt{6}$ 9. $\frac{1}{4}$

11. $7a^3$ 13. $\frac{x\sqrt{10y}}{6y^2}$

8.3 Solving Equations Containing Square Roots

1. $\{127\}$ 3. $\{27\}$ 5. $\left\{\frac{11}{7}\right\}$ 7. $\{10\}$ 9. $\{1\}$

8.4 Applications

1. $7\sqrt{2}$ 3. $13\sqrt{2}$ 5. $\frac{8}{5}$ 7. -5 or -4 9. $8\sqrt{7}$ m

8.5 Higher Roots and Rational Exponents

1. 5 3. $3x^3y^2z^2\sqrt[3]{3y^2z}$ 5. $20\sqrt[6]{7}$ 7. $6\sqrt[4]{2}$ 9. $-(15)^{\frac{1}{4}}$

11. not a real number 13. $a^{\frac{11}{6}}$ 15. $y^{\frac{1}{2}}$

8.6 Radical Functions, Equations Containing Higher Roots

1. $\left(-\infty,\frac{4}{5}\right]$ 3. $\{47\}$ 5. $\left\{\frac{9}{2}\right\}$ 7. $\left\{-\frac{4}{5},0\right\}$ 9. $\{6\}$ 11. 4

Chapter 9 FUNCTIONS

9.1 More on Functions

1. Yes, $D = \{1, 3, 5, 7, 8, 9\}$, $R = \{-2, -1, 0, 3, 4, 5\}$

3. Yes, $D = \{1, 5, 7, 8, 9\}$, $R = \{-2, -1, 3, 4\}$. 5. $\left(\frac{1}{3}, -\frac{7}{3}\right)$ 7. $(6, 22)$

9. $2x^2 - 3x - 2$ 11. $-10a - 5h$ 13. Yes

9.2 The Algebra of Functions

1. $x^2 + x - 1$, $D = (-\infty, \infty)$ 3. $-3x + 1$, $D = (-\infty, \infty)$

5. $\dfrac{x^2 - 3}{2 + x}$, $D = \{x | x \neq -2\}$ 7. $\dfrac{4}{5x - 15}$, $D = \{x | x \neq -3 \text{ and } x \neq 3\}$

9. undefined 11. $28t^2 + 7$, $D = (-\infty, \infty)$ 13. $49t$, $D = (-\infty, \infty)$

9.3 Graphs

1.

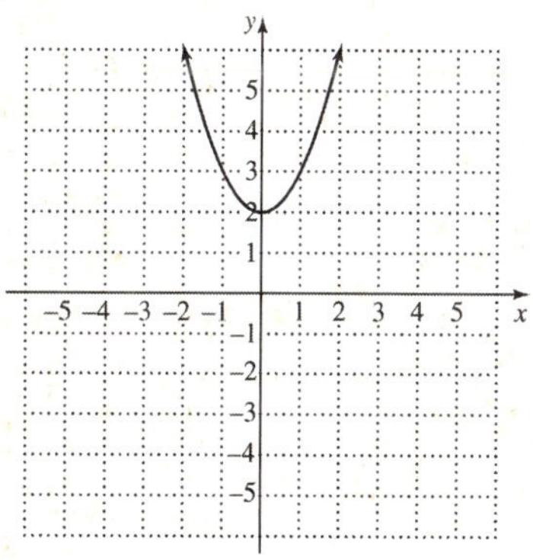

$D = (-\infty, \infty)$
$R = [2, \infty)$

3.

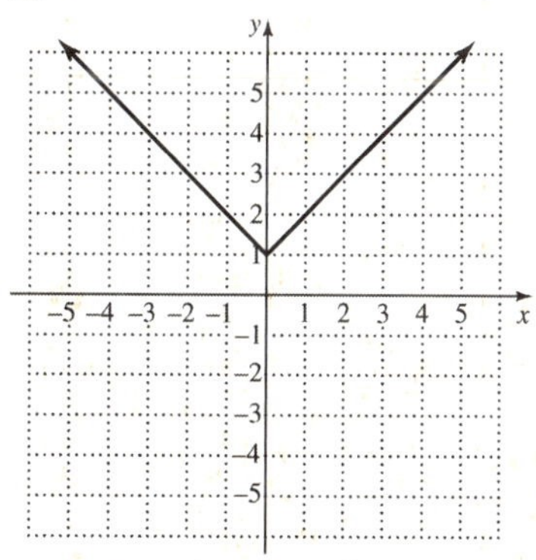

$D = (-\infty, \infty)$
$R = (-\infty, \infty)$

5.

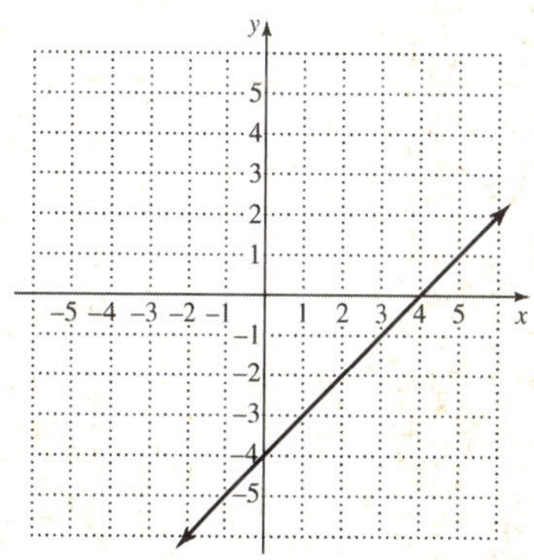

$D = (-\infty, \infty)$
$R = [1, \infty)$

9.4 Translations

1. This is the graph of $f(x) = x^3$ shifted down 1 unit.

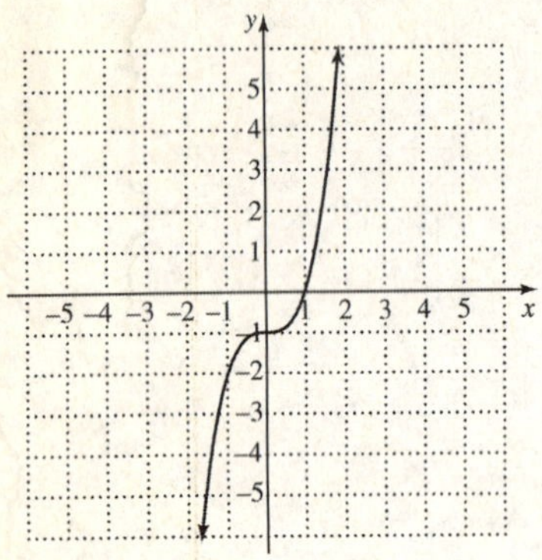

3. This is the graph of $f(x) = x^2$ reflected about the x-axis and shifted up 3 units.

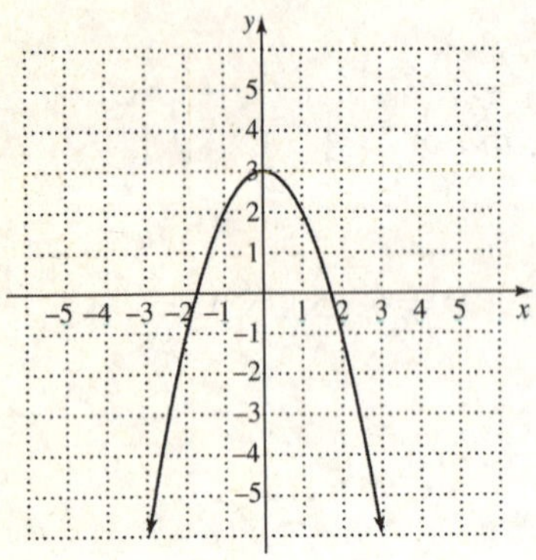

5. This is the graph of $f(x) = x$ reflected about the x-axis and shifted down 3 units.

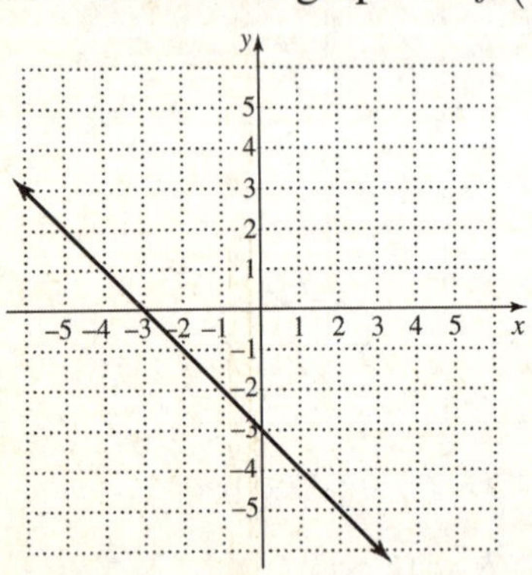

Chapter 10 COMPLEX NUMBERS AND QUADRATIC EQUATIONS

10.1 Complex Numbers

1. $7i$ 3. $\frac{8i}{5}$ 5. $9+3i$ 7. $\frac{3}{5}-\frac{1}{8}i$ 9. $-18-12i$

11. $45+0i$ 13. $6+2i$ 15. $-\frac{1}{17}+\frac{13}{17}i$

10.2 The Square Root Property, Completing the Square, and Quadratic Functions

1. $\left\{-\frac{1}{4},\frac{7}{3}\right\}$ 3. $\left\{-\frac{5}{4},\frac{1}{3}\right\}$ 5. $\{-10i,10i\}$ 7. $\{4,10\}$

9. $\{3+i,3-i\}$ 11. $\frac{49}{36}$ 13. $\{-3-i,-3+i\}$ 15. $\left\{-\frac{3}{2}-\frac{1}{2}i,-\frac{3}{2}+\frac{1}{2}i\right\}$

10.3 Quadratic Equations: The Quadratic Formula

1. $\left\{-\frac{7}{2},3\right\}$ 3. $\left\{-\frac{1}{2}-\frac{5}{2}i,-\frac{1}{2}+\frac{5}{2}i\right\}$ 5. $\left\{\frac{5-\sqrt{73}}{6},\frac{5+\sqrt{73}}{6}\right\}$

7. two complex solutions 9. $\left\{-\frac{5}{8},0\right\}$ 11. $\left\{-\frac{2}{3}-\frac{\sqrt{2}}{6}i,-\frac{2}{3}+\frac{\sqrt{2}}{6}i\right\}$

13. $\left\{4-\sqrt{31},4+\sqrt{31}\right\}$

10.4 Equations Quadratic in Form and Applications

1. $\left\{-i\sqrt{3},i\sqrt{3},-\frac{\sqrt{6}}{6},\frac{\sqrt{6}}{6}\right\}$ 3. $\left\{-\frac{\sqrt{42}}{6}i,\frac{\sqrt{42}}{6}i,-1,1\right\}$ 5. $\left\{-\frac{2}{3},\frac{1}{6}\right\}$

7. $a=\frac{\sqrt{6x}}{3b}$ 9. It would take Huck 9.1 hours and Tom 11.1 hours.

10.5 Quadratic and Rational Inequalities

1. $(-4,3)$;

3. $[0,8]$;

5. $(-\infty,5]\cup[6,\infty)$;

7. $[-8,8]$;

9. $(5,\infty)$;

11. $(-9,3]$;

13. $(-\infty,1)\cup(8,\infty)$; 

Chapter 11 CONIC SECTIONS

11.1 The Parabola

1a. down b. $x=0$ c. $(0,4)$ d. $(0,4)$, $(-\sqrt{2},0)$, $(\sqrt{2},0)$

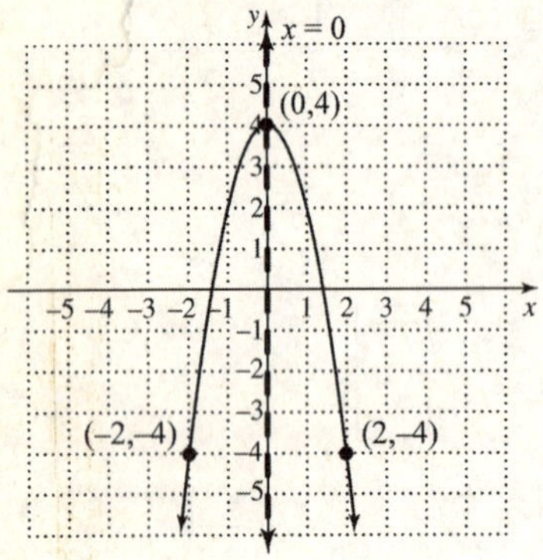

3a. right b. $y=0$ c. $(-3,0)$ d. $(-3,0)$, $(0,-\sqrt{3})$, $(0,\sqrt{3})$

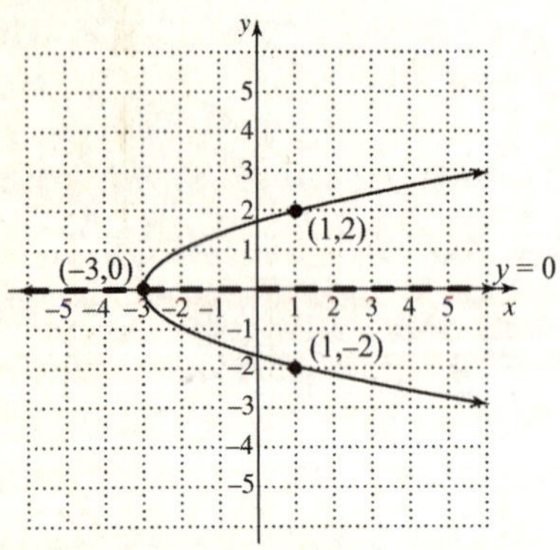

5.

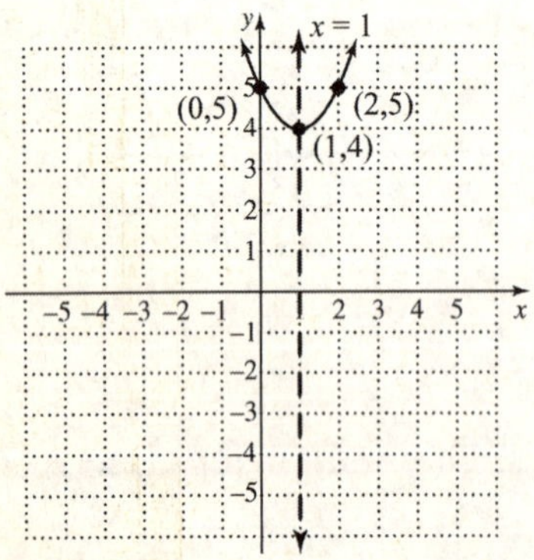

7.

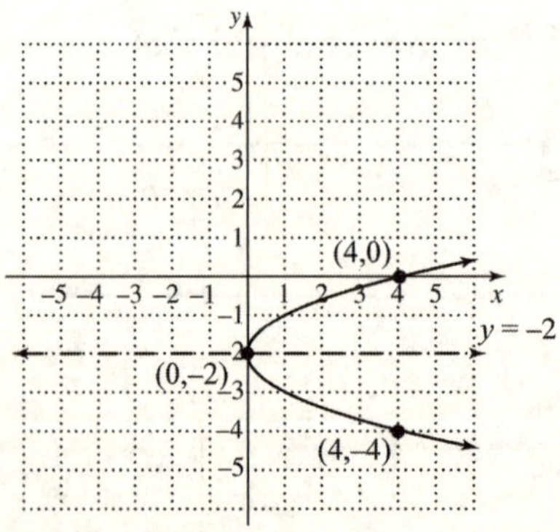

9. $y=(x-6)^2+5$ a. $(6,5)$ b. $x=6$ c. up

11. $x = -3(y-1)^2 + 5$ a. $(5,1)$ b. $y = 1$ c. left

11.2 The Circle and Ellipse

1.

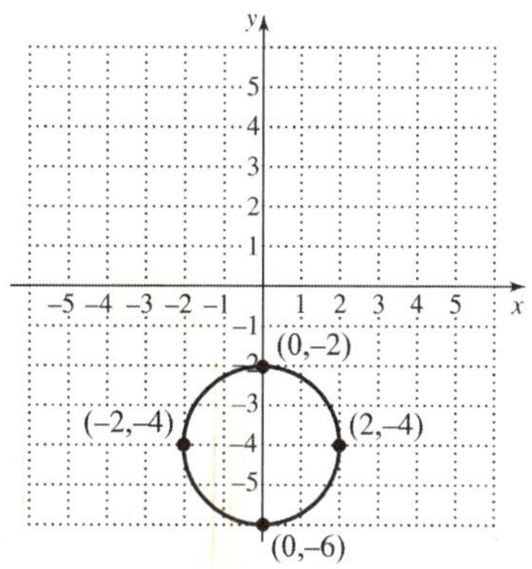

3. $(x+4)^2 + (y-2)^2 = 2$

5. $(x+5)^2 + (y-2)^2 = 16$; radius = 4, $C\ (-5, 2)$

7.

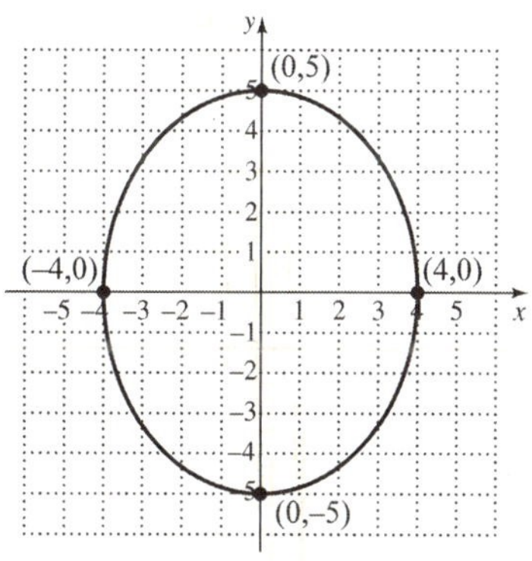

9.

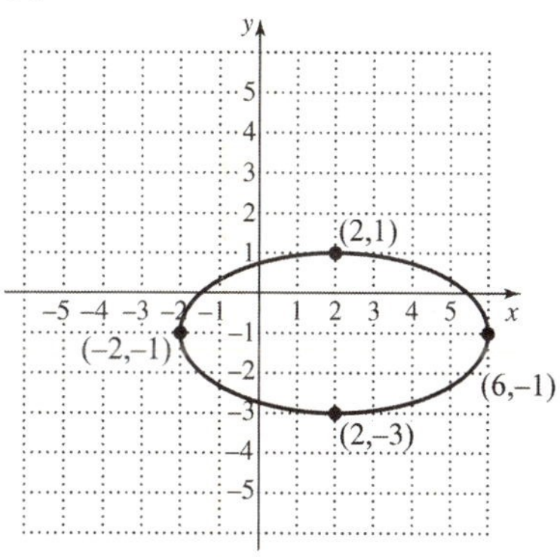

11. $\frac{x^2}{49} + \frac{y^2}{81} = 1$ 13. parabola 15. circle

11.3 The Hyperbola

1.

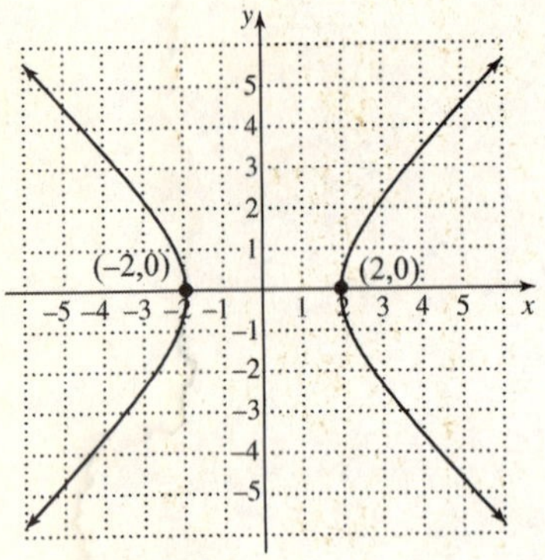

3.

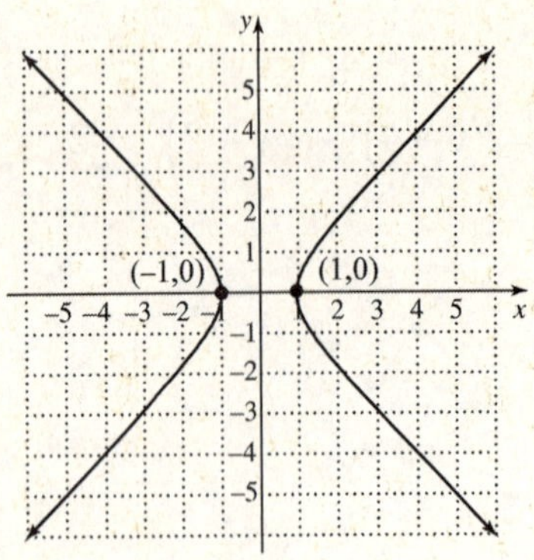

5.

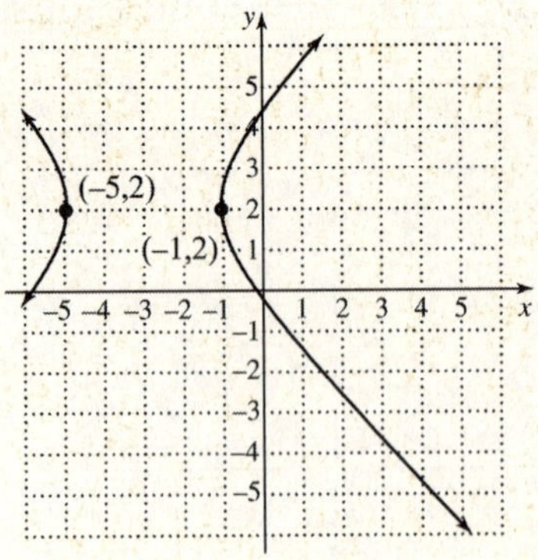

7. $\dfrac{(x+4)^2}{9}-\dfrac{(y-5)^2}{36}=1$

9. hyperbola

11.4 Systems of Nonlinear Equations and Inequalities

1. $\{(3,0),\ (0,-3)\}$

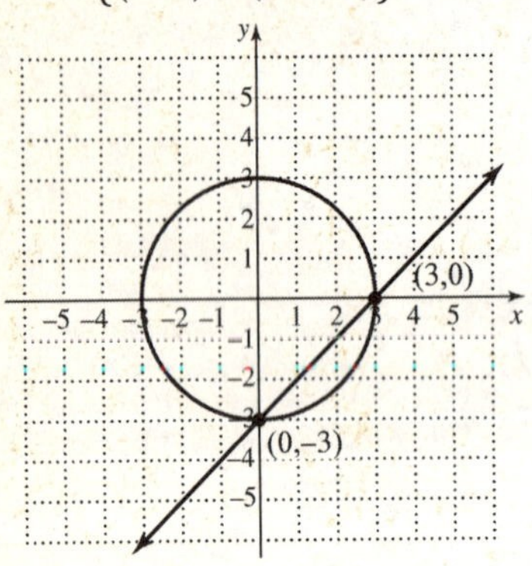

3. $\{(-6,2),\ (4,2)\}$

5. $\{\ \}$

7.

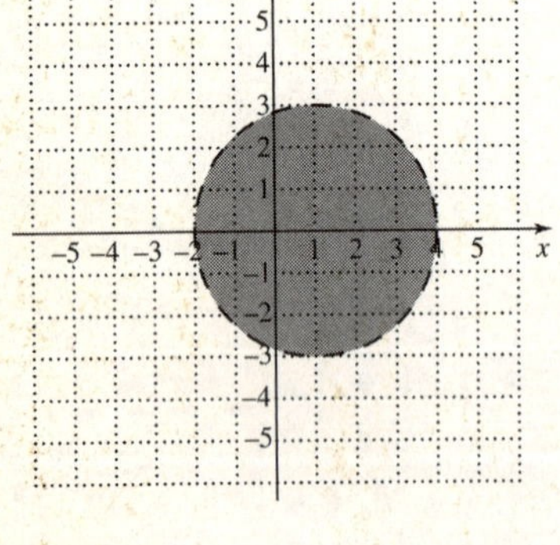

Chapter 12 INVERSE, EXPONENTIAL, AND LOGARITHMIC FUNCTIONS

12.1 Inverse and One-to-One Functions

1. One-to-one, passes the horizontal line test 3. Not one-to-one, fails the horizontal line test 5. One-to-one, passes the horizontal line test 7. yes 9. yes

11. $g^{-1}(x)=x^2+9,\ x\geq 0$ 13. $f^{-1}(x)=\dfrac{10}{x},\ x\neq 0$

12.2 Exponential Functions

1.

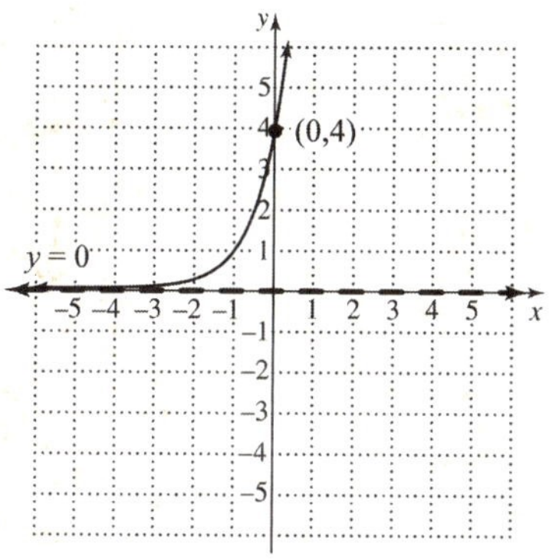

3.

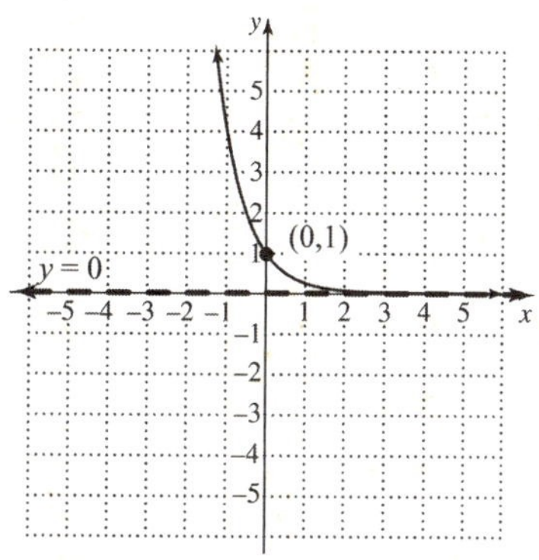

5. 0.945, 2, 2.568 7. $\left\{\dfrac{1}{3}\right\}$ 9. $\{-1\}$

12.3 Logarithmic Functions

1. $4^4=256$ 3. $64^{\frac{1}{3}}=4$ 5. $\log_3\left(\dfrac{1}{27}\right)=-3$ 7. 3 9. $\dfrac{1}{3}$

11. −2.699 13. 6 15. 3

12.4 Properties of Logarithms

1. 1 3. 3 5. $\log_6(8)-\log_6(y)$ 7. $\log_4(7)+\log_4(x)-\log_4(9)$

9. $\ln\left(x^6y\right)$ 11. $\log\left(6561x^4\right)$ 13. $\ln(10)$ 15. 1.14

12.5 Logarithmic and Exponential Equations and Applications

1. 1.4729   3. 9.5330   5. -0.11   7. 3.47   9. $\left\{-\frac{2}{3}\right\}$

11. $\{2,4\}$ 13. Miranda

Appendix B RATIO AND PROPORTION

1. 5 to 7; 5:7; $\frac{5}{7}$ 3. $\frac{1}{8}$ 5. $\frac{3}{10}$ 7. 91   9. 40 centimeters